Routledge Revivals

Regional Economic Development

First published in 1970, *Regional Economic Development: The Federal Role* examines the economic viability of depressed areas and regions. This book questions whether there is adequate justification, on national economic grounds, for Federal involvement with distressed area revitalisation and looks at the main guidelines for intervention in relation to typology of distressed areas. This title is an ideal resource for students interested in politics, macroeconomics, and federal policies concerning the revitalisation of distressed areas and economies.

Regional Economic Development

The Federal Role

Gordon C. Cameron

RFF PRESS
RESOURCES FOR THE FUTURE

First published in 1970
by Resources for the Future, Inc.

This edition first published in 2016 by Routledge
2 Park Square, Milton Park, Abingdon, Oxon, OX14 4RN
and by Routledge
711 Third Avenue, New York, NY 10017

Routledge is an imprint of the Taylor & Francis Group, an informa business

Publisher's Note

The publisher has gone to great lengths to ensure the quality of this reprint but points out that some imperfections in the original copies may be apparent.

Disclaimer

The publisher has made every effort to trace copyright holders and welcomes correspondence from those they have been unable to contact.

A Library of Congress record exists under LC control number: 70142108

ISBN 13: 978-1-138-94476-3 (hbk)
ISBN 13: 978-1-315-67168-0 (ebk)
ISBN 13: 978-1-138-94480-0 (pbk)

REGIONAL ECONOMIC DEVELOPMENT

THE FEDERAL ROLE

GORDON C. CAMERON

Published by Resources for the Future, Inc.

Distributed by The Johns Hopkins Press, Baltimore and London

RESOURCES FOR THE FUTURE, INC.
1755 Massachusetts Avenue, N.W., Washington, D.C. 20036

Resources for the Future is a nonprofit corporation for research and education in the development, conservation, and use of natural resources and the improvement of the quality of the environment. It was established in 1952 with the cooperation of the Ford Foundation. Part of the work of Resources for the Future is carried out by its resident staff; part is supported by grants to universities and other nonprofit organizations. Unless otherwise stated, interpretations and conclusions in RFF publications are those of the authors; the organization takes responsibility for the selection of significant subjects for study, the competence of the researchers, and their freedom of inquiry.

This book is one of RFF's regional studies, which are under the direction of Lowdon Wingo. Gordon C. Cameron, who is senior lecturer in applied economics at the University of Glasgow, began this study while he was a visiting scholar with RFF during 1967. The book was edited and designed by Pauline M. Batchelder.

RFF staff editors: Henry Jarrett, Vera W. Dodds, Nora E. Roots, Tadd Fisher.

Manufactured in the United States of America
Library of Congress Catalog Card Number 70-142108
ISBN 0-8018-1271-2
Price $4.00

ACKNOWLEDGMENTS

That this monograph should have appeared in print is due, in no small measure, to the encouragement and counsel of many friends and colleagues. I am particularly grateful to Harvey Perloff, who originally suggested that this work should be undertaken, and to Bill Miernyk, Dave Bramhall, and Lee Martin, who all made searching and helpful comments on a first draft.

The debt of gratitude to Edgar Dunn and Lowdon Wingo is especially strong—to Edgar for his ready willingness to be disturbed from his own work and for his wise words of encouragement, and to Lowdon for his real contributions to the finished manuscript and his tact and patience in waiting for it!

My contacts with Economic Development Agency officials were always helpful and stimulating, and in this regard John Lindley deserves special mention. Needless to say, none of the abovementioned bears any responsibility for any remaining inaccuracies or errors in the text.

As always, an author is powerless without a long-suffering secretary, and here I was especially lucky to have my draft hieroglyphics made presentable by Shirley Deininger of Resources for the Future and Audrey Graham of Glasgow University.

My last word of thanks is for my wife, without whose constant understanding and encouragement this work would have become a treatise on economic history!

Gordon C. Cameron
University of Glasgow
May 1970

CONTENTS

TABLES

REGIONAL ECONOMIC DEVELOPMENT

THE FEDERAL ROLE

1

INTRODUCTION

It is all too common a practice nowadays to christen as an "economic miracle" every phase of rapid national expansion, but whether this is an appropriate evaluation of the performance of the U.S. economy in the sixties or not, there can be no dispute over the staggering growth in the nation's productive capacity. Between 1960 and 1966 the increase in real gross national product was larger than the total annual production of goods and services in every other country except the U.S.S.R. Over the same period, real disposable income, per capita, rose by 20 percent. Furthermore, in most of these years, growth was achieved under conditions which have come to be regarded as the attainable ideal for mature economies, that is, modest price inflation, limited underutilization of plant capacity, moderate labor-force unemployment, and a higher proportion of GNP allocated to fixed investment in the terminal as compared to the initial year. Thus, all the evidence indicates a highly satisfactory national rate of growth, with a high proportion of capacity being utilized, as efficient adjustments were made to the characteristics and spatial distribution of factor supplies in response to radical changes in demand. Only a persistent disequilibrium in the balance of payments would seem to prevent the administration from being awarded a straight "A" on its performance.

And yet, when the focus of analysis is shifted from the macro to the micro scale, it becomes obvious that national growth and full employment have diminished, but not eradicated, the obstacles to the full employment of distinguishable sectors of the work force. This is most immediately apparent for certain types of worker and particularly for the nonwhite populations, for whom a combination of low educational attainment and discrimination have created a high incidence of unemployment and persistently low median incomes. The factors which cause a high rate of unemployment among teenagers and older employees are more subtle, but probably associated with imperfect knowledge of available opportunities by the unemployed, and perhaps discrimination of a different kind. Thus employers may assume that the marginal product of such workers is less than the total wage and fringe-benefit cost at which these workers are willing to supply their labor.

An even more intractable problem arises from the existence of a large number of areas and even regions in which well below average median incomes prevail, and other once-prosperous localities in which high unemployment persists over a number of years. In such areas the symptoms of economic stagnation or decline swamp the few signs of market-induced structural adjustment. Thus, even including exogenous transfer payments, average returns to local factors of production are significantly below the national average and unemployment, underemployment, and low activity rates indicate substantial reserves of labor. A general environment of depression is sustained by the depreciation of private and social capital, by the poor quality of public services, and by the continuous out-migration of the young and relatively more skilled.

The full extent of this problem is difficult to gauge, but in June 1966, when national unemployment was at 4 percent (the administration's interim target full-employment rate) and GNP was continuing to rise, seven major, fifty medium-sized, and over 400 minor mainland labor markets with a combined labor force of over 2 million were officially designated as areas of "persistent unemployment." At the same date 261 of the 3,000 counties in the country had an annual median family income 50 percent or less of the national median figure and were qualified for special Federal assistance.

For the United States, as for other developed economies, there is nothing unfamiliar in the coincidence of national full employment with persistently high unemployment in some localities. At the peak of every postwar expansion, some once-prosperous areas have contained a surplus of unemployed labor considerably in excess of frictional levels, and there is every reason to suppose that this was true of earlier periods as well. County income statistics also reveal areas and even regions of persistently low income. Nor is it a new phenomenon for the Federal administration and the major political parties to accept as a national problem the persistence of high localized unemployment and spatially extensive poverty during periods of aggregative full employment and to recognize that fiscal and monetary policies with impact on the aggregate levels of demand and supply, and spatially neutral in design, can provide necessary but not sufficient conditions for the elimination of the problem. Between 1949 and 1961, the Truman and Eisenhower administrations introduced limited and uncoordinated measures to assist states and localities which were trying to alleviate personal hardship in the distressed areas and more generally in all areas classified as labor-surplus. Federal defense and other contracting procedures favored such areas, and the Departments of Commerce and Labor provided information and research assistance to help communities attract and develop new employment and identify needed changes in the characteristics of surplus labor. Thus, prior to the first legislation on area redevelopment in 1961,[1] the political debate had begun to shift away from the essentially sterile controversy over whether the Federal government should pursue redevelopment or stimulate outmigration of surplus labor, to a lengthy discussion on the proper balance between these complementary policies and the necessary conditions and policies for generating a process of area redevelopment.

The radical shift in the 1960's is the strength of the forces which favor some form of Federal involvement with area redevelopment. This is apparent both in a series of legislative enactments and in the vigor with which the administration has sought the long-term economic viability of currently depressed areas and regions, through a conscious and carefully considered policy of discriminatory assistance. Here, the overall strategy focuses on three goals: first, to favorably alter the supply characteristics of the depressed communities by investment in public infrastructure; second, to encourage new private investment and induce self-generating growth; and finally, to prevent the reemergence of the depressed area problem by creating joint Federal-multistate institutions with powers to initiate and coordinate regional plans and advise all levels of government on cooperative devices for promoting a satisfactory rate of regional economic growth. To achieve these ends, the Public Works and Economic Development Act of 1965[2] authorized funds for the establishment of one executive agency, the Economic Development Agency,[3] with major responsibility for coordinating and directing Federal activities outlined policies for stimulating area redevelopment, and indicated the scope for further research. Under Title V of the act, Regional Action Planning Commissions have been created in five regions, and as a result of separate legislation a similar

1. The Area Redevelopment Act of 1961.
2. Public Law 89–136, reproduced as Appendix to this book.
3. Hereinafter referred to as EDA.

though distinct regional commission has been formed to deal with the economic development problems of the most distressed region, Appalachia.[4]

Several assumptions underlie this Federal involvement. The first is that many depressed localities and state governments lack the necessary resources to successfully alleviate personal hardship; or in striving to succeed they may misutilize resources in competitive subsidies which encourage transitory private investments having no long-term remedial impact upon the basic weaknesses of the local economy. Accordingly Federal subsidies are seen as the means whereby net employment can be created and personal incomes increased in the distressed areas without the accompanying dangers of permanent subsidization or employment losses in areas where Federal subsidies are not available.

The second assumption is that Federal subsidies to ensure "place prosperity" are justified on national efficiency grounds since they help to create a more optimal spatial distribution of economic activity than that which would result from the operation of the free market. This claim extends far beyond the simple economic case that national output can be raised if subsidies are used to provide employment for the immobile unemployed who currently produce nothing and also require transfer payments. It rests upon the assumption that the depopulation of some distressed communities, and population migration to expanding metropolitan areas, creates social costs which greatly exceed the total gains accruing to the migrants, the new areas in which they settle, and the areas which they have left. In the depressed areas, the returns to immobile local factors of production are reduced and fixed overhead capital is underutilized, whereas in the metropolitan areas fixed capital is duplicated, traffic congestion is increased, and social problems are perpetuated. In time a persistent outflow of capital from the distressed areas denudes local entrepreneurship and creates a real danger of irreversible decline. However, these imperfections in the system are likely to continue just as long as public investment decisions are shaped by arbitrary political boundaries and private investors mistakenly regard slow population growth in the distressed areas as a sign of deficient economic opportunities. In such circumstances any attempt to lower the national rate of unemployment by general fiscal and monetary policies must simply result in even faster growth for the oversize metropolitan areas and therefore even greater social costs.

4. Appalachian Regional Redevelopment Act of 1965.

Consequently, the long-term objective of Federal policy is to mitigate the structural rigidities resulting from imperfect decision making, thereby permitting a greater use of macro policies and an attainment of lower unemployment without inflation.

Whatever the validity of these claims (and there is no general consensus on this), the opportunity to give them substance depends upon a much more limited "proof" of the value of the area redevelopment program. When the existing Public Works and Economic Development legislation expires in 1970, program evaluation will not center on a priori claims but on the substantive evidence that chosen policies have helped to restore some long-term distressed areas to economic health and prevented other communities from lapsing into prolonged distress. All the evidence from developed economies with long-standing programs of area redevelopment suggests that the attainment of such a goal requires considerable skill and originality in the selection, blending, and implementation of national policies and a willingness to provide a framework for constant reevaluation of the overall strategy. Successful area redevelopment requires a total system transformation in which diagnosis and policy prescription must embrace and alter the socio-cultural, as well as the physical-economic, environment. Thus, economic incentives are only one possible method of stimulating compliance with local redevelopment objectives. Moreover, these objectives in part must be determined by local participants, and yet conform to national standards of performance. This requires considerable ingenuity in the setting of such standards so that local initiative is not stifled. Similar considerations arise in the selection of the spatial framework within which the problem is defined and the solution sought. Should distressed areas be defined in terms of explicit characteristics of distress, or larger or smaller areas with political significance? Should distress be seen as an outcome of arbitrary and nonoptimal political boundaries, so that area redevelopment should formulate solutions for new economic or cultural entities? These and other questions of area designation are also related to the determination of program goals. For example, should assistance be directed primarily to the areas of greatest need or to those with maximum potential for economic development? And what socio-economic variables indicate the achievement of program goals in particular designated areas? Without the resolution of this problem, the overall program evaluation and the detailed analysis of particular policies cannot be conducted effectively.

These problems are germane to every national program for area redevelopment, but the U.S. program operates within limits which suggest that the solution of the problems will require particularly sophisticated treatment. The most obvious factor is that Federal involvement must be coordinated with the efforts of states and distressed localities, to prevent diffusion or even contradictions in policy goals. Secondly, the Federal government has no powers over the location choice of private entrepreneurs and cannot use inducements to relocate plants where this would lead to the reduction of employment in the original area. Thus, Federal inducements to attract private development to distressed areas must compete in the open market against the promotional efforts of states and localities without distressed area problems. And finally, the novelty of full Federal involvement and the relative paucity of systematic and independent evaluation of the methodology of area redevelopment mean that no rigid and well-established program guidelines are extant.

Given these conditions, the justification for this monograph is simply stated. Since the area redevelopment program is experimental and voluntaristic, it is likely that the main thrust of EDA policy will be constantly reevaluated and that continual adjustments will be made to the content and emphasis of programs and inducements. This study seeks to aid this process of reevaluation by an objective and detailed study of the policy choices which were made during the period between the passing of the legislation in August 1965 and the end of the Johnson administration (November 1968). Thereafter, the possibilities for developing new policies and new institutions, which might generate self-sufficient growth in some existing or potential distressed areas, will be examined. In this, the distinction first made by John Friedman[5] of *allocative planning*, where the emphasis is on increased and in some sense balanced flows of public capital and finance for problem regions, as contrasted with *innovative planning*, where new institutional arrangements are sought, will be constantly borne in mind.

Several cautionary points are necessary. The political forces which resulted in Federal legislation and currently help to shape the content and execution of Federal regional policy are complex, often hidden, and worthy of study in their own right. A similar statement can be made with regard to the intricate administrative framework

5. *Journal of the American Institute of Planners,* Vol. XXXV (1969), pp. 188–89.

which has been evolved to implement chosen policies. However, this study is concerned with these political and administrative factors only when they illuminate the philosophical and budgetary constraints which limit the policy makers' freedom to innovate.

Secondly, Federal efforts to stimulate regional economic planning and regional development extend far beyond the bailiwick of EDA and its operations. For example, the Department of Defense continues to give preference to contractors in labor-surplus areas provided the submitted bids are competitive. The Office of Education is increasingly concerned with supplementing the resources of poorer states so as to guarantee an improved quality and range of local educational facilities. Efforts by the Department of Labor to provide job information on nearby large urban labor markets for the unemployed and underemployed in poor rural areas are being actively pursued, and of course the Department of Transportation has responsibility for the interstate highway program. However, EDA's originality is that it seeks to generate a *process of economic development in specified areas* of the country. Thus it differs from the other agencies because it examines the development inducing effects of a *range* of public investments and because its spatial focus for investment is defined in advance. In the long run, therefore, one valid test of the EDA success will be whether the various agency programs are given some cohesive purpose, especially when major investments are being considered for individual lagging regions and areas. However, this study will not be concerned with the current regional impacts of other agencies except where there is explicit evidence that EDA is acting as a coordinator of multiagency activity.

Thirdly, the study will concentrate primarily upon the problems of the rural distressed areas, in which low median incomes tend to signal economic malaise. In part this choice is made as a result of the inherent severity of the rural problem, but it is also true that viable solutions for rural decline and depopulation are much more difficult to discern when compared with remedial policies for urban areas. An urban area tends to possess a structure of manufacturing activity, a labor force trained in industrial techniques and used to factory conditions, and spatially concentrated and diverse overhead capital.

Finally, this study will not attempt to provide precise quantitative measures of the impact of EDA policies, since any program which deals in long-term development pay-offs must be judged over a longer period than four years. Therefore, critical evaluations will be limited to chosen directions in policy which appear to have been rewarding or unrewarding.

The Outline of the Study

The study has a simple structure with seven chapters. Chapter 2 discusses the theoretical justification for Federal involvement with area and regional redevelopment, and the main lines of Federal policy for each different type of distressed area. The following chapter develops this last-mentioned theme with particular reference to the role of a growth center policy in ameliorating the social and economic problems of low incomes and population loss in rural areas and mineral resource hinterlands. The fourth chapter, which acts as a preliminary to the fifth, deals explicitly with the legislation of 1965, the current U.S. measures of qualification for distressed area status, the scale of the overall problem and of each different type of distressed area, and the probable future location of problem areas. In the fifth chapter a critical appraisal is made of the policies adopted by EDA between mid-1965 and the end of 1968. A brief outline of the activities of the five Regional Action Planning Commissions follows in Chapter 6, and the findings of both of these last two chapters are welded into an overall appraisal of EDA policy in the concluding chapter.

2

THE CASE FOR FEDERAL INVOLVEMENT

In this chapter two main issues will be discussed: first, the whole question of whether there is adequate justification, on national economic grounds, for Federal involvement with distressed area revitalization will be taken up in the context of two conflicting theories about the likely success of Federal efforts; secondly, the main guidelines for such intervention will be reviewed in relation to a simple typology of distressed areas.[1,2]

The Rationale for Federal Involvement – Two Theories

There is little disagreement in the United States that the existence within the country of hundreds of areas of very low income and of persistently high unemployment is a condition worthy of national concern. The question which is disputed is whether the Federal government ought to make efforts to alter the productive structure of such areas so that they may maintain their level of population, balance their trade with competing regions, and achieve a rate of growth in their per capita incomes which approximates the national rate. There are two quite distinct theories on this. Proponents of the *national demand approach* assert that over the long term the competitive forces of the market do create an optimal spatial distribution of economic activity. Therefore if any area does show persistent symptoms of severe distress this should be interpreted as a clear warning that the nation has a declining need for this particular part of national space. Given such conditions the national government ought to encourage some decline in the absolute level of regional activity, for this is an action which strengthens the *national* economy. The alternative thesis, which can be called the *theory of planned adjustment*, assumes that local economic malaise persists precisely because competitive forces do not create an optimal spatial distribution of economic activity. Thus the lagging regions suffer not only

1. An early draft of this chapter was published in *Regional Studies*, Vol. 2, Dec. 1968, pp. 207–20.
2. The typology will be elaborated in Chapter 4.

because of the internal misutilization of their resources but also because external investors, who are unaware of the favorable opportunities for investment within such areas, continue to pour funds into the overexpanded metropolitan areas within growing regions. Such deficiencies in the market system, it is argued, can be overcome by planning for the adaptation of the supply characteristics of the lagging regions so that they become self-sustaining, retain their population, and attract investment away from the oversize metropolitan areas.

According to the first theory, the impetus to regional growth is provided by national demand factors. If any region, or part of a region, has natural or "created" cost advantages over competing regions in the production and distribution of nationally demanded goods and services, then that region will develop an "export base" and will attract entrepreneurship, private capital, and labor. Over time, such regional specialization will result in internal and external economies of scale and optimal factor mixes, so that regional marginal productivity and regional per capita incomes will be high. Of course, should the demand for regional goods and services fall off, either because of a change in national tastes or because of a decline in the relative competitiveness of the region, then export revenue will be reduced. Falling exports may cause a severe rise in unemployment — a situation which may be compounded as regional producers seek greater efficiency through the substitution of capital for labor. According to this theory, however, labor market disequilibrium cannot persist over the long term since the perfect knowledge of available opportunities by the unemployed and by investors will lead them to maximize their returns. Thus if the region possesses any relative or absolute advantages for the development of new export sectors, then private capital will flow and join locally provided investment funds in the creation of a new export base. Ultimately the unemployed will be reabsorbed. However, if the region does not possess any supply advantages, then capital and labor will flow out to areas of greater opportunity, and the region will phase down to a lower level of economic activity, albeit without heavy unemployment or without major differences in the rewards for given skills as compared to other regions.

Proponents of this theory use this simple model to assert that the failure of the distressed regions to capture sufficient new capital to employ the redundant is clear evidence that such regions do not offer a competitive rate of return. Several reasons are given for this assertion. In part it reflects the view that the alleged advantages of lagging regions, such as the elasticity of labor supply at less than average wage rates in rural areas and the reserves of skilled labor and spare social and economic overhead capital in industrial distressed areas, are not, in practice, real advantages. It is argued that the productivity of rural workers is low and that the skilled industrial workers are likely to possess redundant skills, to be old, and to be averse to retraining. In both situations it is claimed that *efficiency wages* may actually be higher than in the growing regions. Furthermore, although in theory spare public overhead capital can accommodate an increase in private and public economic activity with a decreasing marginal cost per unit of output, in practice, it is argued, much of this spare capacity is either unsuited to the demands of new export sectors or has a very short economic life.

Apart from the spurious nature of these alleged advantages, the major assertion is that lagging regions have positive disadvantages which reduce the rate of return on capital below the level which is competitively acceptable. Typically they are regions with a very limited population, so that the local market is too small to permit the development of plants which can gain internal economies of scale. Furthermore, they tend to be locationally remote from the main centers of population and to have poor transportation and communication facilities, features which impose heavy cost penalties on companies operating there as compared to those in or near to the main population centers. Finally, their limited employment scale and their failure to grow must mean that such regions have not generated external economies of scale and therefore are deficient in such crucial supporting business services as labor placement, advertising, computing, and market research.

A very simple framework for national policy can be drawn from this diagnosis. Every open and competitive economy continually develops thrusts which may put severe pressure upon any region with a specialized structure of economic activity. In some instances, the decline in regional export revenue which follows may be such that labor market disequilibrium is inevitable. The only method of overcoming disequilibrium without heavy net migration of capital and labor is to change the economic (export) base of the region, and this entails capturing part of the plant expansions by leading industrial sectors. However, most lagging regions do not offer any significant advantages to such sectors and, indeed, may legitimately be regarded by them as regions offering a relatively poor rate of return. Furthermore, there is little that local governments can do to overcome regional comparative disadvantages since the national developmental thrusts are far too powerful to be altered by inexperienced administrators with highly limited financial resources at their disposal. Thus if the national government tries to prevent the decline of particular industries through subsidies, then this will reduce the level of regional unemployment and stem the flow of outmigration, but it will also interfere with free factor movement and lock resources into economic environments in which factor returns are less than optimal. In consequence, although a policy of regional subsidization can be justified on the political and social grounds of redistributing income to areas of need, it may actually produce a level of output which is below that which could have been produced without subsidies. Furthermore, the subsidies would have to be permanently offered so as to overcome the persistent disadvantages of lagging regions. Such a policy therefore cannot be justified on the ground of improving national welfare. Indeed, if a policy of maximizing national output is to be followed, then the national government ought to encourage the outmigration of capital and labor from the lagging regions to areas of "natural" growth.

As we have already noted, the theory of planned adjustment adopts the argument that lagging regions are only one manifestation of a national failure to achieve a spatial distribution of economic activity which would maximize national output. Inherent in this approach are three crucial assumptions. The first is that lagging areas are unable to overcome their structural difficulties because increasing amounts of public and private investment are being poured into large metropolitan areas which have expanded far beyond the point which is optimal for the nation as a whole. In part this argument rests on the allegation that at a certain urban population scale, the logistic problems of ensuring that the community is regularly supplied with water, adequate educational facilities, refuse collection services, and so on become virtually insoluble. Indeed in such a context, current production technology and public service management may never be sufficiently proficient to cope with the growing demands of urban producers and consumers so that continual breakdowns in service and sharply rising unit costs become inevitable. However, the critical assertion is that despite these "agglomeration diseconomies," metropolitan areas continue to expand because many important production, investment, and consumption decisions are based almost entirely on the knowledge of the private rather than the social costs inherent in the decision.[3]

Of course, it is true that part of the social costs generated by private activities are internalized by the private decision maker and therefore should, in theory, choke off the rampant expansion of the community and its activities. The private decision maker cannot totally isolate himself from the effects of traffic congestion, air and water pollution, and noise, and in any event will face higher land costs and increased tax burdens brought about by the steeply rising costs of providing essential public services. And yet these restraints may be insufficient in so far as the decision process reflects only part of the social costs which ultimately will be borne directly by the decision maker. This, it is said, is seen most clearly in the case of migrants who, in moving to the metropolitan area, may increase their earnings sufficiently to cover their short-term migration costs, and still provide net long-term benefits. However, this movement may become excessive precisely because migration is subsidized by existing metropolitan residents who make tax payments which more than cover the costs of the public services consumed by them, whereas the migrants make especially heavy marginal demands for these services and yet pay local taxes on an average cost basis.

Clearly this discrepancy between private costs and benefits and social costs and benefits can occur in every urban area of any given scale. Yet proponents of this theory argue not only that the level and variety of unwanted externalities rise sharply as the urban community expands beyond a certain point, but also that there can not be any satisfactory public solutions without a rigorous control over metropolitan growth. For example, to expand the capacity of urban services is inevitably a self-defeating expedient. Hansen has explained this view in the following terms:

> ". . . short-run, cost-benefit comparisons may favor the expenditure of a given amount of highway funds in a congested urban area rather than in an alternative region because, even though cost per mile will be higher in the former, marginal short-run benefits will be still greater. However, the new highway's contribution to already existing agglomeration economies will serve to attract new

3. This means that the unwanted social externalities are not internalized by the decision unit in question; and of course this decision unit may be a public corporation as well as a private individual.

population and economic activity to the congested area. Thus the long-run effect will be to increase congestion and associated social costs; more people will live under the same state of congestion as the original project was intended to relieve."[4]

Furthermore, doubts are cast on the feasibility of charging private individuals and corporate bodies the difference between the private and the social costs of their activities, either because no such precise figure could be delineated or because such a measure would be income-regressive in its impact. It follows that if the growth of very large concentrations of population and economic activity inevitably creates insuperable managerial problems in the provision of basic urban services, and uncontrollable social costs, then the marginal public investment costs of servicing an increment to national output would be distinctly lower in lagging areas where some social and economic overhead capital is underutilized and new capital could be provided without attendant internal and external diseconomies of scale.

The second key argument is that certain types of firms could actually reduce their costs of operation by decentralizing from the congested metropolitan areas. In this view, the major metropolitan areas ought to be the locations for large-scale, complex-cycle activities which draw their inputs from a wide variety of urban sources and for companies involved in developing new products and new production technologies in which initial "breakthrough costs" can be minimized by easy access to a large pool of highly skilled labor and specialist business and research activities. Such areas inevitably become the high cost areas, not only because of the diseconomies of scale mentioned previously and the competition among major companies for urban sites but also because high income economies consume an ever increasing amount and variety of labor-intensive services in which the chances of productivity gains are minimal. Given these conditions and the relative decline in importance of transport and communication costs, the lagging areas present a low-cost environment for production branches which use simple production techniques, have limited material and human inputs, and require relatively limited management control from head plant level. Once again, however, the assertion is that the free market does not guarantee such a distribution of economic activity since management is unaware of, or prejudiced against, the real merits of the low-cost, lagging regions.

4. Niles M. Hansen, "Towards a New Approach in Regional Economic Policy," *Land Economics*, Vol. XLIII (1967), p. 377.

The final assumption is that lagging regions by their own efforts can win an increasing share of private investment over the long term provided they are given short-term assistance. Thus regional and community success in redevelopment is a function of the willingness to probe deeply into any weakness of the local economy, the degree of sophistication in identifying, attracting, and stimulating the development of new employment sectors which can draw upon regional economic advantages, and the adequacy of planning in providing the correct sequence and oversight for remedial investments which facilitate the smooth implantation of these new sectors. Now since the key principles of this process can be readily assimilated, it follows that many distressed regions and communities are capable of learning how to devise strategies for structural adaptation and, with the expert help and financial involvement of the national government, of implementing these strategies successfully.

Entirely different parameters for national involvement with regional adjustment follow this whole diagnosis. Thus subsidies for lagging regions are justified on grounds which are extremely similar to those for "infant industry" protection. The assumption is that the lagging regions can become viable and provide sufficient job openings for growing regional populations provided national subsidies are used to ease them through their period of transition and to prevent a cumulative loss of resource advantages. This transitional assistance must be geared to overcoming any competitive disadvantages which the region possesses and to the creation of net advantages. If this policy is properly formulated then lagging regions can greatly improve their competitive efficiency by "internal" action. But they can also provide a sound economic environment which limits outmigration to, and attracts private investment from, the oversize metropolitan areas, where the public costs of servicing growth are exorbitant and the social costs inherent in uncontrolled and unwanted externalities ever more apparent. On two counts then, improved resource utilization within the regions and a more optimal spatial distribution of national economic activity, subsidization for lagging regions can be justified on the grounds of national efficiency rather than income distribution.

The Case for Subsidies

The evidence which is available suggests that neither of these theories provides a sufficiently comprehensive framework for national involvement with the redevelopment of lagging regions nor for a national strategy on the total distribution of population and economic activity. However, both theories make unique and significant contributions to our understanding of why local structural difficulties persist in specific conditions, and both clear some of the ground necessary for the formulation of this national strategy.

The great merit of the demand theory is its stress on the variety and power of the dynamic factors at work in every competitive economy. These developmental "thrusts" continually modify regional competitive advantages so that regional specialization, which may bring high income rewards in periods of above average marginal productivity growth, also can provide the conditions for long periods of structural "shock," labor market disequilibrium, and low per capita income growth. Broadly defined, these development thrusts can be classified into basic and longstanding structural modifications to the economy as a whole, which affect all regions of the country, though not necessarily with equal force, and secondly, the more limited developments within a given economic structure which have a limited impact upon specific employment sectors and possibly upon specific locations. As regards the first type of structural modification, the U.S. economy is continuing to move into the stage where net value added and employment are dominated by the service sector, where manufacturing activity is relatively less important and primary activity is in relative decline. Four major "change factors" are at the root of this process. The first is the high income elasticity of demand for services in which output gains are realized mainly through absolute increases in labor inputs, rather than through productivity increases.[5] The second is the high income elasticity of demand for those manufactured goods produced in complex-cycle processes and reliant upon highly processed inputs and particularly light metals, ferroalloys, fabricated construction materials, and petroleum products. Thirdly, the process reflects a long-run tendency for the demand for most primary products to rise much less than the growth in per capita incomes. In particular, agricultural output has risen moderately on the basis of a massive and constant reduction in the labor force and the substitution of mechanized equipment, fertilizers, and scientific farm management. Finally, the changes embrace a long-run and continuing process of resource substitution which is especially noticeable in the fossil fuel sector where bitumi-

5. See, for example, the article by W. J. Baumol, "Economics of Unbalanced Growth," *American Economic Review*, Vol. LVII (1967), pp. 415 ff.

nous and anthracite coal are increasingly being replaced by petroleum and natural gas.

The impact of these major developments on the national economy has been dramatic. Less than forty years ago (1929) over a quarter of the U.S. labor force was employed in agricultural and mining activities. The current figure is less than ten percent. In contrast, the proportion of employees in service and construction activities was less than half in 1929 but now stands at almost two-thirds. This trend will continue in the future and probably at an increased rate. Indeed, a recent projection of labor demand suggests that of the 16 million additional jobs which may be created between 1965 and 1975, almost 80 percent will be in the services sector.[6]

This structural modification of the total economy has hit some regions with especial severity. Regions which once supported large agricultural populations have suffered continual and sizable reductions in the demand for labor as a result of a shift in activity to more profitable land elsewhere and/or because of a substitution of capital for labor. Economic activity in coal-mining towns, which developed in response to the demands of the manufacturers in New England and Middle Atlantic states, has dwindled as these demands have fallen or because of mineral extinction. Certain areas which specialized in simple types of manufacturing have faced constant reductions in the demand for labor due to the outmigration of manufacturing activity.

In addition to these major and irrevocable thrusts of the economy, the export base of every region is under constant pressure from more limited competitive and political forces which are largely generated outside of the region. For example, export revenue may decline if there is finished product or natural resource substitution, a fall-off in government orders, or the removal of tariff barriers or quotas on foreign goods. Alternatively, exports may become uncompetitive because of the discovery and exploitation of a new natural resource in another region, the relative inability to control labor cost increases, or a relative failure to invest in cost-cutting fixed investment and organizational improvements.

The effects on particular local economies of these exogenous pressures may be a drastic reduction in particular employment sectors. Indeed, the evidence suggests that many major depressed areas have experienced staggering reductions in exports and employment

6. U.S. Department of Labor, *Manpower Report of the President,* 1966.

within a decade. For example, for many of the 14 standard metropolitan statistical areas (SMSA's) which were depressed in all or most of the years of the fifties, employment declines in basic sectors of over 45 percent were not uncommon (see Table 1).

Table 1. Employment Decline in Basic Sectors, 1950–60

State	SMSA[a]	Major sector of decline	Percentage of total employment		Percentage decline in sector employment, 1950–60
			1950	1960	
Indiana	Evansville	Motor vehicles	16	9	47
"	Terre Haute	Railroads	6	3	49
Massachusetts	Fall River	Textiles	25	14	43
"	Lawrence	Textiles	36	5	65
"	Lowell	Textiles	24	7	70
New Jersey	Atlantic City	Other personal services	15	11	9
North Carolina	Durham	Textiles	9	6	26
Pennsylvania	Altoona	Railroads	32	19	44
"	Erie	Nonelectrical machinery	11	5	56
"	Johnstown	Mining	23	8	60
"	Scranton	Mining	13	4	76
"	Wilkes-Barre	Mining	23	5	81
Rhode Island	Providence	Textiles	18	9	49
West Virginia	Charleston	Mining	20	4	49

[a] SMSA = Standard metropolitan statistical area.

The demand theory also is correct in stressing that the root problem of the distressed areas is an uncompetitive structure of economic activity and that the only chance of long-term viability for such areas is to develop a revised economic base. A regional solution which relies on shoring up uncompetitive industries must remain as a permanent "solution" and cannot be phased out. If subsidies are withdrawn, then in time the payments out of the regional economy for imported goods and services, for external capital investments, and for national taxes will be consistently larger than the payments received into the local economy from exports, capital inflows, transfer payments, asset sales, and exogenous investments, so that there will be an automatic reduction in local expenditure and local activity.

Finally, there is abundant evidence to suggest that labor does migrate away from areas of relatively poor economic opportunity. A detailed analysis of migration flows into and out of distressed areas, conducted by Eva Mueller and associates,[7] showed that areas of relatively low income and of heavy unemployment tended to export more population than they gained by inmigration whereas the reverse was true of areas of better economic opportunity. Moreover, as a large proportion of outmigrants were young adults of child-bearing age, the total long term effect on population growth within the losing areas was considerably greater than the net migration figures indicated.

And yet, despite the realism and authenticity of the demand theory in these several respects, there is overwhelming evidence to suggest that the self-equilibrating process of factor adjustment is by no means an entirely valid model of what happens to regions and areas which are faced by the prospects of persistent disequilibrium. What we can show is that despite national full employment, geographical immobilities in local labor markets have created a condition in which (i) national output cannot be increased without subsidies for those areas where immobilities occur; (ii) subsidies for those areas are likely to generate a greater increase in national real output than if such subsidies were used in areas operating at the margin of labor capacity.

Both these propositions are validated by the fact that the income-maximizing theory of factor mobility rests upon an extremely simplistic explanation of human behavior under conditions of labor market disequilibrium. What we know now is that such an explanation takes us only so far in explaining actual phenomena. For example, the data on gross migration flows to and from distressed areas reveals a very complex pattern. Certainly the areas of low median incomes tend to lose proportionately more population through outmigration than the richer areas. But the really poor counties do not have proportionately greater outflows and may even have lower outmigration than the better-off counties within the overall low-income grouping. This suggests that where there are marked inter-regional income disparities, population will flow from the poor areas to the richer, but it does not follow that the really poor areas will have sufficiently large outflows to offset their very unfavorable posi-

7. E. Mueller, N. Barth, and W. Ladd, *Migration into and out of Depressed Areas* (U.S. Department of Commerce, Area Redevelopment Administration, Sept. 1964).

tion. A similar statement can be made with respect to levels of unemployment, since there is no evidence to suggest that the higher the unemployment the greater the outmigration.[8] Indeed, a fascinating study by Lowry of the migration flows of two small SMSA's, one prosperous and gaining population, the other one with lower median incomes and losing population, has shown that the volume of outmigration was not a function of absolute or relative levels of economic opportunity, but of the size and structural properties of the resident population, and particularly its age distribution.[9]

Where the depressed areas really differ is in their lower rate of inmigration and in the disparity between the average productivity of the outmigrants and that of the inmigrants to the distressed area. Between 1950 and 1962, 20 percent of all 1962 residents moved into areas of high and persistent unemployment, 22 percent into areas of low income, and no less than 31 percent into all other, nondistressed areas.[10]

Part of the explanation for a level of outmigration from depressed areas which does not lower labor supply to match existing labor demand may be extremely poor knowledge of labor market conditions elsewhere. In the study by Mueller, it was found that almost half of those questioned in depressed areas thought that there was an equal or greater amount of work locally available as compared to other areas. What is even more surprising is that almost 60 percent of the respondents in the areas of chronically low median income thought that local rates of pay were higher than or the same as those elsewhere.[11] But better information may not be totally effective since for some groups the result of calculating the relative benefits and costs of migration and of remaining unemployed or underemployed will always be to stay put. Older workers, persons with low educational attainment, and those with strong family and social ties may consider that the certainty of a transfer income and the security of a familiar environment are more valuable than the uncertain benefits following geographical movement. Alternatively, the unwillingness to migrate may be related to past investments in business or private houses, which, with a decline in economic activity, yield lower returns, and can only be sold in falling markets at

8. Mueller, Barth, Ladd, pp. 8–10.

9. Ira S. Lowry, *Migration and Metropolitan Growth* (University of California, Los Angeles, 1966), pp. 93–98.

10. Lowry, p. 11.

11. Lowry, p. 10.

prices which do not provide sufficient capital for a new start elsewhere. Of course, in some instances low outmigration may simply reflect the unwillingness of the unemployed to recognize that a further employment decline is inevitable or the failure of local and state governments to defray the costs of migration.

Whatever the causes, persistent job scarcity may generate underemployment of labor. For example, Miernyk has described how organized job-sharing by textile workers may take the place of permanent unemployment.

"Under a program of job-sharing two workers in effect hold down a single job. While one is employed his alternate is drawing unemployment compensation. After some interval the employed worker is laid off and begins to draw compensation while his alternate returns to work in the mill. Thus . . . a total of 5000 textile jobs may be 'held' by approximately 10,000 workers."[12]

There is some evidence that persistently depressed areas have lower activity rates for men. A study of the comparative changes in the rates of twelve major depressed areas and all other urban areas between 1950 and 1960 suggested that in such areas the young and the old do not have the same social and economic pressures to participate as workers in the "prime earning years," and are therefore willing to avail themselves of higher education opportunities or early retirement.

In conditions of major structural shock, then, the level of unsubsidized outmigration is unlikely to be sufficient to permit a restoration of equilibrium in the labor market, except perhaps over the very long term. Moreover, the national decision as to whether to intervene with assistance for any particular area must be made at a very early age of structural decline. There are two reasons for this. The first is that in the absence of outside assistance the internal capacity for successful adaptation may progressively decline as high quality labor migrates out and is replaced by labor of lower proficiency which joins an aging stock of largely immobile workers. The second reason is that in the face of persistent outmigration of high quality labor, local government may be forced into the defensive strategy of freezing tax levels so as to prevent any loss of industry. This approach not only results in a depreciation of physical assets and public utilities at precisely the period when prospective inves-

12. W. Miernyk, "Depressed Industrial Areas—A National Problem" (National Planning Association, Jan. 1957), p. 17.

tors are making a close scrutiny of the area's potential but it may also result in an underinvestment in education, which generates a net flow of poorly educated workers with low mobility potential.[13] This, of course, relates to a wider point about the proficiency of local governments in devising the correct strategy to restore equilibrium. Most communities affected by a major structural shock engage in promotional subsidization. There are conflicting views on whether such subsidies improve national welfare, but whatever their merits on these grounds, there is substantial evidence to suggest that local promotional efforts as a whole are largely ineffectual in overcoming persistent disequilibrium. It follows that the redevelopment potential of most distressed regions and communities cannot be effectively measured by the success, over the short term, of local governments in altering the productive structure of the local economy. Thus, some communities may contain considerable reserves of human and overhead capital and yet fail to attract sufficient exogenous investment.

We are now in a position to distill the essence of these two theories.

1. Every open economy with specialized regional structures can expect considerable variations by region in the rate of growth of employment.
2. If for any reason, in a given region, export revenue falls in an important sector, then heavy unemployment, underemployment, and perhaps wage differentials for given skills may emerge.
3. In such conditions the market mechanism will automatically redistribute some labor and some capital to more productive activities outside of the region.
4. The unsubsidized outmigration of labor is unlikely to be a sufficiently elastic response to the decline in local employment opportunities. The result is that serious unemployment, underemployment, and wage differentials may persist.
5. The unsubsidized outmigration of labor is unlikely to be a totally satisfactory solution to local persistent disequilibrium for one or more reasons.

13. The justification for subsidies to increase the level of investment in education is entirely different from the justification for shoring up declining industries. In the latter case the long-term demand for the industry's products is downward so that the life of fixed assets in use is limited. A failure to invest in education may reduce the earnings prospects of labor for as much as 50 years of the "asset's" life.

(a) In regions in which the long-term chances of attracting exogenous capital and of stimulating indigenous expansion of economic activity are not high, then the net loss of population due to an excess of outmigration over inmigration is unlikely to be sufficient to restore equilibrium.

(b) In regions in which the potential for redevelopment is uncertain, but the marginal social infrastructure cost of raising output is low, the net loss of the skilled, young, and enterprising may reduce the chances of attracting exogenous investment which could restore equilibrium.

(c) In regions in which the redevelopment potential is high over the long term, the net loss of the highly proficient sectors of the labor force may reduce the chances of attracting the most advanced industries.

6. The scale and rapidity of the external "shocks" to regional structures may be such that local governments and promotional agencies are incapable of devising strategies to restore equilibrium. Their defensive actions may, in fact, result in an underinvestment in education which generates a flow of educationally handicapped workers.

These points suggest that in the short run there may be four valid economic grounds for national subsidies to lagging areas. Thus where labor is unemployed in the lagging regions and cannot be induced to move to areas of greater opportunity, then subsidies which result in the creation of jobs and net output will increase national output. Even if the subsidies were used in the prosperous full-employment regions where the unsubsidized private rate of return is higher, they would increase output by an amount which was less than if they were used in lagging areas. This situation arises because the subsidized increase in output in the prosperous region is equal to the incremental gross output minus the opportunity costs of labor, whereas in the lagging areas unemployed labor has no opportunity costs, so that the subsidized increase in output is the gross output raised.[14] Secondly, there may be lagging regions with utilizable and underused social and economic overhead capital. If such conditions exist, an unfavorable private rate of return on capital could be offset by the low marginal public investment cost of raising output, so that the social (public and private) incremental

14. Clearly national subsidies which create employment opportunities also reduce the volume of transfer payments to the unemployed.

capital-output ratio might be highly favorable. Once again, national subsidies to encourage private investment in the lagging area might be justified on the grounds of maximizing real output.[15] Thirdly, underinvestment in education by local governments creates the necessity for the national government to provide grants to prevent a potential flow of low productivity workers. Finally, there may be a social case for buttressing older workers against the inevitable material and psychological costs of rapid structural decline. This may take the form of the subsidized creation of new work opportunities, lump-sum early-retirement benefits, or enhanced unemployment benefits.

Area Redevelopment Subsidies in the Long Run

Although there are sound grounds for subsidizing lagging regions in the short run, the critical question is what should be the long-term objective of subsidization. As we have already noted, this depends upon the assumptions made about the developmental potential of such areas and the rationale for permitting the unimpeded expansion of fast-growing metropolitan areas. In both of these respects the theories under review are unhelpful because they are doctrinaire. Thus the logic of the first theory is that governments can do nothing to alter the spatial distribution of economic activity, whereas the extreme position of the other view is that governments can guarantee satisfactory redevelopment in every lagging community and can know precisely what overall spatial distribution of economic activity is optimal.

Part of this conflict over the possibilities of regional redevelopment arises from a rigid adherence to one specific delineation of what is meant by successful redevelopment. Broadly defined, the demand school assume that successful redevelopment must imply a reduction in the volume of *regional activity* and probably in the regional population. In contrast the planned adjustment adherents visualize that *each lagging community* can maintain its level of population and achieve a structural transformation which guarantees a satisfactory growth in productivity and in per capita incomes. However, there are a number of optional states between these extremes which could legitimately form the basis of a satisfactory

15. Thus stress on real output is important. Regions with sparse human and fixed capital can be expected to accommodate an increase in demand for local resources without inflation.

objective function for national involvement. We will return to this point in later chapters. Secondly, both theories make the mistake of estimating developmental potential from only one part of the total time framework necessary for this estimation. The demand theory, in focusing upon the period of shock, assumes that a historical failure to alter the regions' productive structure is indicative of a future inability to alter that structure. The adjustment theory tends to ignore the severity of the shock and the reasons for lack of adaptation in the past, and stresses how development is largely a learning process which can be stimulated in almost any environment. And yet, it is just as invalid for the first theory to concentrate on prediction from historical data on growth, while ignoring the possibilities that one exogenous investment decision can alter an area's prospects, as it is for the other to overstress those development thrusts which wholly destroy the rationale for certain types of location. At this stage, then, what we must do is engage in a general discussion of the key variables which seem to affect regional and area redevelopment potential and thereafter delineate the precise form of subsidies, both short run and long run, which should be advanced to each particular type of lagging area.

An assessment of redevelopment potential must contain a historical review of the reasons why the local economic base ossified, why the current employment shock was particularly severe, and why normal market forces have not solved the crises. We shall deal with each of these points in turn.

1. A priori, it is reasonable to expect that the length of the period of structural adjustment will depend, in part, upon the nature of the decline in export activity which initiated the period of disequilibrium. E. A. G. Robinson has shown that this decline can vary considerably in its severity:

> "If the region is large and the decline of exports is confined to a single relatively small industry, the net loss of local incomes will be relatively small in proportion to total income, the possibilities of absorption of those displaced in other occupations exporting from the region will be high, the ratio of imports to all expenditures will be lower. If the region is small, undiversified and greatly dependent upon a single activity, the possibilities of absorption into other exporting activities will be low, the proportionate decline of all incomes will be greater, and since the region is small and specialised, the ratio of imports to all expendi-

tures is likely to be greater; thus the speed and ease of adjustment are likely to be less. . . ."[16]

There are three essential points contained in Robinson's diagnosis. The first relates to the scale of the direct decline in export activity, the second to the interrelatedness of local sectors with the declining export activity, and the third to the absorptive capacity of other growing export sectors. Other factors may be important. For example, the rapidity of the decline in the key sector, the early recognition of the seriousness and permanence of decline, and the extent to which the decline is caused by an outmigration of plants and entrepreneurship to competing regions, may all affect the potential for redevelopment.

2. It is impossible to summarize adequately the major locational trends which may enhance an area's potential or make it highly suspect, but several factors appear to be of major importance. First, the range of locational choice open to the manufacturer has widened enormously in the last few decades. Fewer and fewer industries are dependent upon locations close to the source of their raw materials, and the vastly improved systems of interstate transport and communications have permitted the establishment of large productive units which gain internal economies of scale, thereafter transmitting their output to regional and national markets at relatively low cost. Secondly, there is evidence of increasing homogeneity in production costs, and particularly wage costs, throughout broad areas of the continent. Furthermore, technological knowledge seems to flow without impediment between the major urban centers. Thirdly, the relative decline in importance of these purely economic factors has initiated an increasing number of location decisions on the basis of the quality of the local environment and of its climate. All these factors suggest that, provided the national government uses sensitive fiscal and monetary weapons to balance the aggregate level of demand with the full productive capacity of the nation, then every *region* may be capable of sustaining a satisfactory rate of growth in employment, and some which are especially well endowed with natural and environmental features have particularly favorable opportunities for rapid expansion. This thesis tends to be confirmed both by the marked convergence in the ratio of manufacturing employment to population among the various regions of the

16. E. A. G. Robinson, ed., *Backward Areas in Advanced Countries: Proceedings of a Conference Held by the International Economics Association at Varenna* (New York: St. Martin's Press, 1969), p. 17.

Table 2. Growth in Nonagricultural Payroll Employees by Region, 1947–66

Region	Number (× 1,000)		Increase (× 1,000)	% Distribution		Average annual growth rate		
	1947	1966	1947–66	1947	1966[a]	1947–61	1961–65	1965–66
All regions	43,443	63,070	19,627	100.0	100.0			
New England	3,333	4,156	823	7.7	6.6	0.8	1.9	3.6
Middle Atlantic	10,813	13,015	2,202	24.9	20.6	0.6	1.8	2.6
East North Central	10,067	13,381	3,314	23.2	21.2	0.9	3.1	4.2
West North Central	3,414	4,780	1,366	7.9	7.6	1.5	2.4	4.0
South Atlantic	5,269	8,941	3,672	12.1	14.2	2.3	4.1	4.8
East South Central	2,148	3,392	1,244	4.9	5.4	1.8	4.0	5.0
West South Central	3,059	5,131	2,072	7.0	8.1	2.4	3.5	4.4
Mountain	1,170	2,264	1,095	2.7	3.6	3.6	3.0	4.3
Pacific	4,170	8,009	3,839	9.6	12.7	3.3	3.6	5.9
Average annual growth rate						1.6	2.9	4.2
Mean deviation						1.0	0.7	0.7
As % of average						62.5	24.1	16.7

Note: From U.S. Department of Labor, *Manpower Report of the President,* Apr. 1967, Table 5, p. 26.

[a] Provisional.

country[17] and by the small deviation in regional employment growth rates from the national growth rate, during the period of full employment and fast expansion in the 1960's. The latter point is shown clearly in Table 2.

The trends in per capita income are less clear cut than those in employment. However, even here there is evidence that many of the poorest states have improved their relative position in recent years.[18] Using per capita personal income data for each state as a ratio of the national income figure for 1965, 1966, and 1967, and the provisional figure for 1968, we have looked at income growth performance over these years. As Table 3 shows, nearly all the states which started with incomes more than 20 percent below the national figure have grown at a faster rate than the nation and therefore converged. With states which had 1965 incomes up to 20 percent lower than the nation's, the pattern is much less clear cut, with 8 growing faster, 11 growing more slowly, and 4 showing the same rate of growth as the nation.

Table 3. State Income Growth in Relation to Nation, 1965–68

1965 ratio of per capita personal income (Nation = 100)	Number of states[a]		
	Converging	Diverging	No change
50 – 59.9	1	0	0
60 – 69.9	2	1	0
70 – 79.9	5	0	1
80 – 89.9	6	6	1
90 – 99.9	2	5	3
100 – 109.9	3	2	2
110 – 119.9	5	2	0
120 +	2	1	1
Total	26	17	8

[a] Includes the 50 states and the District of Columbia.

With the states which had above average incomes in 1965, the pattern of growth is much more clear cut, since more than half

17. See, for example, the introduction to B. Chinitz, ed., *City and Suburb: The Economics of Metropolitan Growth* (Prentice-Hall, 1964).

18. The author wishes to thank Mr. J. D. McCallum for permitting him to use the following calculations, which are drawn from Mr. McCallum's ongoing thesis on regional economic planning in the United Kingdom and the United States. His data source is the *World Almanac*, 1970, which drew its figures from the Office of Business Economics, U.S. Department of Commerce.

grew more slowly than the nation whereas only approximately 30 percent grew at a rate above that of the nation.

3. Yet the crucial point is that growth is not spread evenly within each region. All the evidence suggests that in the competition for capital and human resources, the larger urban areas have marked advantages over smaller communities. A few statistics will make this clear. Twenty-four metropolitan areas had populations of half a million or more in 1960. In 1900 less than one-fourth of the U.S. population lived in these places. The 1960 figure is over one-third, and if current trends continue about half of the U.S. population will live in these two dozen places by the end of the century. On a finer level of spatial disaggregation, the trends are even more startling. Between 1950 and 1960, when the largest decennial increase in population in the nation's history was recorded (28 million), almost half of the 3,110 county units in the United States[19] actually *lost* population. These losses were concentrated in the smallest units. In 1950, the population living in counties of 50,000 inhabitants and above was 104 million, that for counties of less than this size, 47 million. Ten years later, the absolute growth in population in the larger counties was 31 million, but the smaller counties actually experienced an absolute decline of 3 million.

Thus, with notable exceptions, the efforts of small urban areas and rural counties to stem structural decline and to prevent continuous outmigration have been signally unsuccessful. In 1966, 885 counties out of a total of 3,134 in the United States were qualified by the Economic Development Administration as being distressed either on grounds of heavy unemployment or because of low income and heavy outmigration or low income alone. Significantly, the proportionate qualifications were far heavier among the smaller counties of under 50,000 inhabitants than for the larger counties.[20] Moreover, as we will show in a later chapter, the small labor markets have responded least well to the fast growth of the national economy over the last few years. In 1963, 528 areas were qualified as distressed on grounds of their substantial and persistent unemployment. Four years later, when the rate of national unemployment was steady at around 4 percent, only one-third of the large and medium-size labor markets remained qualified. However, at the same date almost 57 percent of the smallest size group remained qualified. In general, then, high redevelopment potential seems to be a func-

19. I.e., excluding county units in Alaska.
20. See Table 4.

tion of the size of the community, so that the prospects for small communities distant from major urban centers appear to be distinctly unfavorable.

4. The deep-lying and intricate causes of a process involving growth or persistent decline cannot be summarized in a few words, but six factors are of obvious relevance. The relative decline in the importance of raw material inputs for most industries, the growing use of semifinished manufactured inputs and subcontracted services, and the desire to minimize costs of access to final markets have all contributed to a preference for location within, or close to, the major urban centers of activity. This has been further enhanced by the growing demands of innovating production units for those unique inputs which are found only in large centers, especially business consultancy, advertising, marketing, and computing services, the large pools of highly trained labor, and the unique urban communication channels through which are radiated critical technical and business data. It must also be recognized that an increasing number of the export industries are in the service rather than the manufacturing sector, and in many of these the median size of company is very large, so that proximity to a large and varied pool of labor is essential. Finally, although research on the causes of migration is inadequate, it appears that it is the range of high income opportunities and the adequacy of social services which "pull" the rural poor to the cities rather than the "push" of their intolerable rural conditions.

5. There is no reason to suppose that the economic advantages of the medium and large urban areas will be diminished in the years ahead. Indeed, a recent projection to 1975 of growth in employment by size of county suggests that three-quarters of all the counties in the population size range from 50,000 to 500,000 will grow faster than the national average rate. However, this projection also suggests that the bulk of the smallest counties (1–50,000) and the largest counties (over 500,000) will fail to match the national growth rate.[21]

The Different Types of Distressed Area – Revival or Decline?

Given this information, we can now attempt to evaluate the scope for long-term involvement with area revival in each type of dis-

21. U.S. Department of Commerce, EDA, Apr. 1967. See Chapter 5 for a further discussion of the likely scale of distressed area problems in the future.

tressed area. In a later chapter we will describe and analyze the socio-economic and geographic characteristics of several different types of distressed area.[22] Meanwhile it is sufficient to list four basic types, that is, fast growing, *high income,* but high unemployment areas; *old manufacturing* areas, with heavy unemployment, relatively high per capita incomes, and net outmigration; the *not-so-poor rural* areas suffering from heavy unemployment; and finally the *rural* areas where exceptionally low incomes and sometimes very high unemployment result in heavy outmigration.

The long-term solutions for the areas of fast growth but labor surplus are fairly clear cut, and almost certainly realizable without substantial assistance by the Federal government. Information campaigns directed at producers facing severe labor shortages in other centers within the region, backed up by low-interest loans and training supplements, together with improved labor market guidance and migration allowances for the unemployed, might simultaneously increase the demand for, and reduce the supply of, labor in these rich, surplus-labor areas.

In many of the second type of area (the rich, old industrial area) there is abundant evidence to suggest that there is an internal motivation and organizational ability to recreate the structural fabric. Here Federal assistance should be broad-gauge and supplement existing internal efforts to improve the functioning of local capital markets, freight and personnel communication systems, educational systems, labor and management training programs, and subregional planning efforts.

In the not-so-poor rural areas, the basic problem is not that the export structure of the economy is weak but that there are inadequate urban centers which can provide basic services to the scattered rural population and provide a wide range of job opportunities for those who cannot, or do not desire to, find employment in agriculture or forestry. One possible solution to this problem is Federal planning assistance to identify service centers and infrastructure grants to achieve a phased build-up of such centers.

The really difficult Federal decisions relate to the solutions for the poor mountain and rural communities. These are the really small and scattered settlements, containing poorly educated and badly serviced residents, employed in low-productivity industries which are likely to demand less and less labor in aggregate and more and

22. See Chapter 5.

more highly trained specialists if they are to remain competitive. The chances of reviving these resource hinterlands are negligible, and every effort should be made to encourage outmigration. The key question is whether these outward migration flows should be deliberately channeled in any specific direction. It is, of course, the assumed connection between the decline of this type of economic area and the alleged overexpansion of the metropolitan areas because of the subsidization of rural migrants which has led some to argue that these large areas should be contained and the depressed rural and mountain communities revitalized.

The traditional case for containment on economic grounds is by no means conclusive. If the migrant flow is excessive because of subsidies then this is a short-term phenomenon, since residential qualification and tax burdens typically can be avoided only in the short run. In any event, it is not at all apparent that inmigration would be substantially diminished by immediate marginal-cost taxing systems. Furthermore, the arguments on the grounds of increased congestion are unconvincing, since it is the proportion of social capital which is congested that matters and there is no evidence that measured in this way smaller communities have a significantly lighter congestion load. It is true that large cities tend to be high wage locations, but whether this leads to a suboptimal distribution of resources must be judged *after* estimates of the incremental gains in productivity from urban scale and accessibility have been made. In fact, the major uncertainty is not whether the major cities are inefficient locations but whether they can absorb large infusions of poorly educated migrants without serious social and political disruption.

Certainly there is abundant evidence that many of the rural poor who migrate to the city ghettoes suffer long periods of unemployment or underemployment largely because most employment growth has been generated at some distance from the central city ghettoes and because deficient transit systems and discrimination in the sale and leasing of suburban housing have provided effective barriers to a match between suburban vacancies and the ghetto unemployed. There is no reason to suppose that spatial disequilibrium of this kind will not be overcome in time by some combination of subsidies to transit authorities to provide central city - suburban travel, by antidiscrimination laws regarding the sale and leasing of suburban housing, by massive training programs for the ghetto poor, and by efforts to provide industrial centers on the fringe of central business districts.

Moreover, since the volume of migration from rural areas to urban areas has declined in recent years, one might assume reduced problems of absorption in the major receiving cities.[23]

However, even with reduced migratory flows, the remaining short-run socio-political problems of absorption cannot but be severe, and thus there remains the question of whether the Federal government should attempt to divert this flow to regional centers within the problem regions.

There is overwhelming evidence that any attempt to divert a large part of this migratory flow is destined to failure unless massive subsidization of regional center growth is undertaken. Thus the lure of the "freedom trail," the high welfare benefits, the long-term hope of relatively high earning job prospects, all of these act as powerful magnets pulling migrants to the major northern and western cities. Therefore, it is unlikely that migration to these cities can, or should, be restrained to any degree. It is conceivable, however, that the rate of return on Federal attempts to divert a small part of the flow may be high if two conditions are present. If any regional centers have spare social and economic overhead then they may be able to absorb an immigrant flow far more easily than those major centers operating at the limits of capacity. Secondly, some migrants who are loath to sever social, personal, or economic ties with their original rural location may prefer a short-distance move to the regional center, which permits the fuller continuation of these linkages. On both grounds, therefore, there may be a case for using some Federal funds, which would have been allocated to the major cities receiving migrants, for these regional reception centers instead.

Of course there are other arguments which can be used to justify this approach – arguments which are unrelated to the alleged benefits of preventing inefficiency or socio-political malaise in fast-expanding cities outside of the problem regions. For example, any policy of regional subsidization which accelerates the long-run

23. Alonso has calculated that only 22.6% of the population growth of metropolitan areas between 1960 and 1966 came from inmigration as compared to 35% in the 1950–60 period. Moreover the bulk of this migration was international in origin rather than domestic, only 6% of overall population growth in metropolitan areas being caused by this domestic migration. Furthermore the vast majority of domestic migrants to metropolitan areas came from smaller urban areas and not from the farms. W. Alonso, "What Are New Towns For?" *Urban Studies*, Vol. 7, No. 1 (1970).

tendency to concentration of a regional population within a few key centers is likely to create the conditions for servicing net and replacement demand for social/economic overhead capital at a low per capita cost. Moreover, for any given amount of subsidy, spatially concentrated investment is likely to have the maximum regional income multiplier effect in the short run; to attract the maximum flow of exogenous capital and enterprise to the region; and to enhance the inherent production advantages of a large urban base, so that new enterprise flourishes and the regional balance-of-payments deficit is reduced. Finally a rapid development of job opportunities in such centers may lure high quality labor into the region from competing regions, thus offsetting to some degree the persistent losses of high productivity labor.

All these points will be examined in detail in the next chapter.

Summary

1. In the short run there are nationally valid grounds for subsidizing the creation of jobs, providing help with local education, and supplementing unemployment benefits in lagging areas.

2. Over the longer run the Federal government must distinguish between those areas in which future viability is made possible or highly impossible by the nature and scale of structural decline, the current national developmental thrusts, and the organized willingness of the local agents.

3. The most significant developmental thrust is the decline of small rural and urban areas in the resource hinterlands and the growing concentration of population in urban areas, and particularly those with more than 50,000 population.

4. The trends suggest that a very large share of the nation's economic activity and population shortly will be concentrated in approximately twenty major urban centers.

5. The larger urban areas as a whole present unique opportunities for efficient utilization of resources and for positive social interaction. The traditional economic arguments for containing the growth of these cities are inadequate.

6. The small number of large urban areas at present suffering persistent disequilibrium have a high potential for future viability without substantial Federal assistance.

7. The medium-size industrial areas within the old manufacturing belt and within the fast growing regions could achieve viability pro-

vided the Federal government assists with consultancy services to identify structural weaknesses and with grants for management and labor training and retraining.

8. The solution to the adjustment problems in the not-so-poor rural areas may be a growth-point strategy which results in the deliberate build-up of selected medium-sized urban service areas.

9. The solution to the really critical socio-economic problems of decline in the poor rural and mountain communities, and the accompanying political and social stresses inherent in large-scale absorption of migrants in northern and western cities, may lie partly in Federal efforts to develop existing medium-sized and large cities *within* the regions of concentrated decline.

3

GROWTH CENTERS AND THE AREAS OF SEVERE DISTRESS

A previous chapter has suggested that the chances of revitalizing the economic structure of many of the really distressed areas in the agricultural and mineral resource hinterlands are negligible given the scale of employment declines in basic sectors and the paucity of the remedial resources at the disposal of Federal, state, and local agencies. Redevelopment planning for such areas cannot proceed from projections of growth based on the existing distribution of regional population. Instead, the precondition for public attempts to establish a new regional economic base must be the deliberate initiation of an entirely new pattern of population distribution. One variant of this approach is a policy of spatial selectivity, in which a small number of "growth centers" are selected for large-scale public investment. The core concept in this approach is that there must be official encouragement of spatially unbalanced growth within a region in the short run, if the level of regional economic activity and regional share of net national value added are to be substantially upgraded over the long run. Put simply, the growth center approach represents an attempt to simulate the growth characteristics of prosperous centers in fast-growing regions with the overall objectives of attracting exogenous capital, of encouraging local product and production innovation, and ultimately of spreading economic improvements to the non-growth center parts of the region. In time this combination of major public investment and effort for selected locations should, it is claimed, result in an alleviation of the regional balance-of-payments disequilibrium. It should also help in the process of substituting intraregional movements for interregional movements of capital and labor which presently threaten to denude the lagging region of critical resources.[1]

Now since economic viability is thought to be correlated with a relatively large urban scale, this process of growth center selection

1. Some of the ideas discussed in this chapter were contained in the author's "Growth Areas, Growth Centres and Regional Conversion," *Scottish Journal of Political Economy,* Vol. XVII (1970), pp. 19–38.

typically focuses upon towns and cities high up in the rank-order of settlement size, and it may either concentrate upon urban centers which have grown rapidly without any special public assistance or alternatively may include centers with a poor growth performance in the past but diagnosed potential for rapid future expansion.[2] Thus growth center policy, which is selective and biased towards a small number of large urban centers, may either accept a determinist view of economic development and "bank upon certainties" or alternatively use public investments to enlarge the number of key regional centers. In simple terms this can be viewed either as an attempt to induce greater economic growth in the existing growth centers or greater economic growth in existing centers *and* in chosen locations. But in either case the spatial distribution of economic growth is different from that which could be expected from a context which lacked deliberate public investment planning.

An obvious concomitant of this policy of concentration is that national and state agencies should do nothing to blur the focus of selectivity and, in particular, should ignore claims for assistance from smaller centers seeking revitalization. Of course, once again, there is considerable latitude here for negotiation over "just claims" for public investment and for different interpretations of how much freedom each community ought to have when pursuing its own economic development.

While all of this may seem self-evident, there is considerable doubt over the proven as distinct from the alleged socio-economic benefits of a growth center policy. Furthermore, little is known of how a growth center policy strategy should be integrated into an overall framework for particular distressed regions and subregions. This chapter will analyze the potential benefits of this approach and will discuss the key policy issues which may arise when attempts are made to implement the strategy.

The Alleged Benefits of a Growth Center Policy

THE MARGINAL SOCIAL INFRASTRUCTURE COST OF DEVELOPMENT

Proponents of the growth center approach argue that it has several distinct advantages over any policy which scatters investments and planning efforts in response to political demands, on the basis

2. A third alternative is the creation of growth centers ("new towns") on virgin sites. Though this is a popular method in European countries and increasingly in the fast-growing regions of the U.S., it will not be considered here.

of need, or for simple per capita requirements. One of the most common claims is that in the long run the net demands for social and economic overhead capital by any regional population can be supplied more cheaply if that population is mainly concentrated in urban centers of considerable size. Thus, whether the regional population is growing rapidly, is static, or is declining, it is assumed that a growth center approach, biased as it is towards the growth of relatively large centers, will contribute to a regional population distribution which minimizes infrastructural costs.

This claim assumes that there is an optimum range of city size within which the marginal infrastructure costs of supporting a given increase in economic activity are at their lowest. Typically this optimal size range is thought to lie somewhere between the largest cities and the smallest centers, so that the marginal costs of urban development have the familiar U-shaped distribution when plotted against increasing size of community. It follows that if the precise limits of this optimal size range can be discerned, then the *cheapest* way to support private growth would be to encourage development in locations within this size range.

Obviously, in any particular region, this simple relationship between the marginal costs of development and community size may be obscured by other factors. For example, unit building costs may vary within the region because of variations in efficiency within the regional construction industry, because of a varying degree of demand pressure upon scarce resources, or indeed because of topographical and structural thresholds which affect the costs of development in specific towns and cities.[3] It also matters whether the anticipated increment in economic activity is on a large scale or not. If it is a small-scale expansion, then the cheapest place to develop might be in communities *of any size* which have infrastructural capacity to spare. Leaving such complications aside we can return to the arguments against providing infrastructure in relatively small centers. Here the crucial assumption is that unit costs of infrastructure fall as scale is increased.[4] For example, the per capita cost of building a public sewer system may fall with an increase in the population served simply because the capacity of the pumping station and

3. See, for example, J. T. Hughes and J. Kozlowski, "Threshold Analysis – An Economic Tool for Town and Regional Planning," *Urban Studies*, Vol. 5 (1968), pp. 132–43.

4. We will not repeat the arguments against development in large cities which were covered in Chapter 2.

sewerage pipes can be increased without a comparable increase in the cost of the materials and labor used in fabrication. Similarly there may be economies in providing large communities with those items of infrastructure which must satisfy a range of unstandardized demands. The classic example is the primary road network, which must be of sufficient width and have the necessary load-bearing qualities to accommodate heavy commercial vehicles, as well as more "common" types of traffic. As a result, design thresholds, which affect the form and scale of the asset, are made necessary by the demands of a limited set of the total users. However, once these assets are provided in an indivisible form, they may be able to accommodate a large and intensive flow of "special traffic" without any increase in maintenance costs. The greater the volume flow of this special traffic, the lower the capital plus maintenance cost per unit journey. Thus the primary route network may cost precisely the same in the small community as in the large, but be used by a very small number of special demanders so that the capital cost per special trip is markedly higher. It is also reasonable to expect economies in the actual construction/installation of many items of social/economic overhead capital. For example, the unit cost of building a large batch of new houses may be significantly lower than that for a small batch, not only because legal charges and the costs of preparing and formally submitting plans may not vary directly with output but also because the large batch may permit large-scale ordering of components and a higher labor efficiency on the construction work itself. Finally, there may be savings on operating cost, as distinct from capital cost, and in this respect large power-generating facilities may have reduced labor and fuel inputs per unit of output.

The published evidence on the shape of the marginal cost curve for social and economic overhead capital as towns increase in size is scanty, unstandardized, and conflicting. Four different types of research inquiry have been conducted. Cross-section studies in the United States have analyzed the variations in total expenditure per head on a range of public services as city population is increased.[5] The obvious difficulty with this approach is that it does not distinguish between capital costs and current costs, nor does it establish to what extent a higher expenditure is caused by a greater

5. A good example is Harvey E. Brazer, "City Expenditures in the United States," National Bureau of Economic Research, Inc., Occasional Paper No. 66 (New York, 1959).

volume of service or a higher quality of service. Some studies have focused solely on capital expenditure, but once again such an approach cannot distinguish between demand and supply factors.[6]

A more useful approach has recently been developed in Italy, where there has been an inquiry into the per capita costs of providing a *standard range of infrastructure assets* for communities of varying size.[7] This study confirms that up to a certain population size there is an inverse relationship between the per capita costs of net infrastructure provision and a growth in population. Indeed, the figures suggest that the infrastructural cost curve conforms to the typical U-shaped distribution, with the minimum cost range lying between communities of 30,000 and 250,000 population. The really significant upward thrust in costs appears to occur in the very small communities of under 5,000 inhabitants and in the large cities with more than a quarter of a million inhabitants.

The obvious conclusion to be drawn from this evidence is that for any given regional population, the pattern of settlement which would minimize the cost of providing essential infrastructure to accommodate a growth in economic activity would be based on communities with more than 30,000 and fewer than 250,000 inhabitants.[8] However, for a variety of reasons this conclusion cannot be accepted without question. Besides those difficulties which are part and parcel of any study of this kind, such as the hypothetical estimation of the capital and operating costs of assets never before constructed on the "required" scale, there were other deficiencies of a less defensible kind. For two major items of infrastructure for which there would, in all probability, be marked capital savings with increased size (hospitals and secondary schools), it was assumed that per capita costs did not vary with community size. Similarly no account was taken of possible economies in installing large items of infrastructure. Both of these omissions mean that the cost advantages of increased scale were minimized, so that the limiting thresholds of 30,000 and 250,000 should probably have been increased. In

6. For development of this point see Niles Hansen, "The Structure and Determinants of Local Public Investment Expenditures," *Review of Economics and Statistics*, Vol. XLVII (1965), pp. 150–62.

7. Associazione per lo sviluppo dell' industria nel Mezzogiorno (Rome), "Ricerca sui coste d'insediamento," 1967.

8. This, of course, assumes that the costs of transmitting goods and people within the region and between the centers and other regional population concentrations do not vary significantly whatever regional pattern of population distribution is chosen.

contrast, land costs were not included in the calculations, so that the advantages of large scale were probably magnified.

The most rigorous piece of research yet completed relates to five Indian cities.[9] Unlike the Italian study, this exercise was conducted against actual urban conditions, the basic idea being to calculate the public investment cost of a given increase in industrial activity in cities which ranged in size from 48,000 to 1,070,000. The calculations took account of surplus capacity in existing public facilities, city land area and form, income levels, and costs of construction labor and materials.

Three results from this study are especially important. Urban infrastructure costs are shown to be relatively insignificant. Thus for every £100 of net value added in manufacturing, the incremental infrastructure cost is of the order of £13.[10] Moreover, the variation in incremental cost over the range of city size was extremely small, but it did show a consistent pattern of decline as city size increased. It represented 14.5 percent of the related value added in the city of 48,000, 13.3 percent in the city of 132,000, and 12.8 percent in the city of 1,070,000. Clearly the differential between the smallest and largest city was small, and between the 132,000 city and the 1 million plus city the unit cost difference was insignificant. The bulk of associated infrastructure provision was in "social" categories (housing, education, hospitals, telephones, fire and police protection, refuse disposal), which accounted for approximately 75 percent. In this category, costs fell irregularly but consistently from the 48,000 city level to the 212,000, increased again for the 323,000 city, and thereafter remained constant. The remaining 25 percent of infrastructure costs are directly associated with industrial development. Here, unit costs fell consistently, though in small amounts, throughout the size range.

How can we interpret this data? We have no U.S. studies, and the most comprehensive European study has major deficiencies. The rigorous Indian study cannot be accepted as the basis of policies in developed countries without some misgivings. Furthermore there is no unqualified agreement on the low and high population size thresholds. Nevertheless it does seem likely that, in most circumstances, unit costs of infrastructure fall as town size increases, and

9. "Cost of Urban Infrastructure for Industry as Related to City Size in Developing Countries" (India Case Study) (Stanford Research Institute Calif., S.P.A. New Delhi, S.I.E.T. Hyderabad, Oct. 1968).

10. An (infrastructural) capital/output ratio of 1:8.

it can also be concluded that the very small urban units of less than 30–40,000 population are serviced at a higher cost than towns in excess of this size range.

CONCENTRATED INVESTMENT AND REGIONAL ECONOMIC GROWTH

Another critical assumption of the growth center strategy is that for any given volume of public investment a policy of urban concentration will result in the largest amount of regional economic growth. Hoover has summarized this argument as follows:

> "... economic improvement initiated in the growth centers will spread to their less-urbanised hinterlands; so that the best way to help these hinterlands in the long-run is not by direct assistance but indirectly by fostering the progress of nearby growth centers."[11]

Hoover has also noted that it is seldom made clear precisely how these economic improvements are likely to be brought about. The argument seems to contain three ingredients. The first is that in the short run a region can gain the maximum secondary or "multiplier" benefits by concentrating public investment within key centers. The second is that concentrated planned investment will attract more exogenous capital and entrepreneurship to the region over the long term than would a series of random and spatially scattered investments. And the third is that concentrated investment will generate an economic environment in the chosen centers which will make them continuously attractive to outside investors and responsive to new opportunities within the national economy. Thus, over the long run these development centers will generate sufficient new job opportunities, at nationally competitive wage rates, to suck in many of the regionally unemployed and underemployed from the hinterlands and to radiate growth impulses to the whole region so that there is work *within* the region for net additions to the labor supply. Each of these arguments must be discussed separately prior to any consideration of the whole thesis.

Initial Multiplier Benefits

At the root of the argument on initial multiplier benefits is the notion of "backward linkage" between urban centers and their economic hinterlands, the latter supplying the centers with inputs

11. E. M. Hoover, "Some Old and New Issues in Regional Development," in Robinson, *Backward Areas in Advanced Countries.*

and finished goods for use in production and in consumption. The centers represent the major market outlets for producers within the hinterland, whereas urban producers are not dependent to the same extent upon hinterland markets. As a result the hinterland areas are always likely to run a favorable balance of trade with the urban centers. It follows that any public investment in the centers will automatically create income there which in turn results in demands to producers in the hinterlands—the secondary effects being far greater in this direction than if public investments were first made in the hinterlands.

It is not immediately apparent why the regional income multiplier effects of a given investment will be greatest if that investment is concentrated within the region's urban centers. Theoretically, the secondary change in regional income from a given injection of public investment, wherever it is made, is some function of the marginal propensity to pay nonregional taxes, the marginal propensity to save, and the marginal propensity to consume regionally produced goods and services. We can assume that the tax liability differences between hinterland income earners and urban income earners are insignificant and therefore should be discounted. Differences in the marginal propensity to save may be more important. With relatively high median incomes in the centers the marginal propensity to save may also be relatively high, so that the income leak is correspondingly higher. On this ground investment in the hinterlands may result in smaller leaks and, other things being equal, a high-income generation effect. However, the crucial question is the extent to which there is any difference between earners in the hinterland and center in their marginal propensity to consume regional products. These differences center upon possible variations in the income elasticity of demand and the extent to which net local demands can be satisfied by regional producers.

Extremely little is known about how marginal consumption patterns vary as between these two types of area, and the best we can do is to make three simplifying assumptions which will facilitate the argument. At a crude level of disaggregation, consumption expenditures can be divided into three types—primary (food and fuel), secondary (manufactured goods), and tertiary (services of all kinds). We can assume that the bulk of the primary and tertiary demands can be satisfied by producers within the region, whereas secondary producers are likely to be highly specialized by region; so that home region suppliers can only satisfy a limited range of the home region's demand. Moreover, in the primary and tertiary sectors the regional net value added is likely to be significantly higher than in the secondary sector, which is not labor extensive and draws a large proportion of its nonlabor inputs from outside of the region. Finally, we can assume that with the generally lower median incomes in the hinterland, marginal additions to income there would be spent in ways different from the marginal expenditures by urban earners. In particular, primary expenditures may be marginally higher, secondary expenditures significantly higher, and outlays on tertiary services which are brought from urban centers distinctly less.

Given these assumptions, comparative expenditure patterns following increments to income might show hinterland earners with relatively large expenditures on primary goods in which there is a high regional net value added, significantly larger secondary expenditures in which there are large leaks outside of the region, and proportionately less outlay on tertiary services, for which there is a high regional net value added. The really important difference between hinterland earners and urban earners might be the high income elasticity of demand for manufactured (secondary) goods by hinterland earners—a demand which results in major income leaks outside of the region.

Therefore there is a strong possibility that the regional multiplier effects flowing from public investments within the urban centers will be greater than the multiplier effects of investing in hinterland areas. Of course, this conclusion is absolutely dependent upon the assumptions made, which may or may not apply to particular circumstances. At best, then, this argument is a feasible, though not a proven, justification for a growth center investment policy.

The Attraction of Exogenous Capital

The growth center strategy often has been proposed as a certain method of minimizing the subsidy cost of attracting private capital and enterprise from prosperous areas.[12] This is based on the simple assertion that carefully chosen centers can provide real opportunities for low-cost expansion whether they receive subsidies or not. Indeed, if subsidies are advanced the assumption is that they will serve to extend existing opportunities so that chosen centers become especially attractive to outside investors. By contrast the inherent attractions of smaller centers are assumed to be markedly less.

12. An alternative way of stating this claim is that for any given scale and mix of subsidies, a growth center policy will maximize the amount of new enterprise attracted to the region.

Once again there is no empirical backing for the claim, but there are four a priori arguments which suggest that it may be valid. By simple virtue of the scale of its consumer purchasing power and width and variety of its economic activities, a large urban area is likely to offer highly favorable market opportunities which attract a continual flow of new companies. Over time, the size and accessibility of this local market may permit such companies to expand their operations and capture internal scale economies in production and distribution. Therefore any policy which seeks to attract enterprise to a series of small centers which, in aggregate, have the same size of population as one or two large centers would not be so successful, precisely because distribution costs and the costs of personal contact with customers would inevitably be greater.

The surpassing of particular scale thresholds may also confer other cost-saving opportunities distinct from those of market scale and the minimization of transfer costs. An airport from which regular air services are operated, a railway freight terminal, a direct road link with an interstate highway, a technical college which provides specific training for local employment conditions, all of these may appear when the urban area is large enough to present favorable opportunities for profitable public or private investment or politically important enough to gain national or state subsidies. Whatever their origins, these investments may present *unique* cost advantages for local producers and local consumers alike which are foregone by enterprises and consumers in smaller urban areas. Loosely, these external economies, which are available to the whole community, may be called the economies of urban scale.

Other external economies may be associated with the scale of the urban labor market. Even if every local labor market had the same rate of unemployment, the large markets would always have the greatest *absolute* amounts of labor and the widest range of skills and experience to offer. In addition, the large area is likely to possess more developed facilities, both public and private, for the placement and training of surplus labor. In combination these factors may result in distinct advantages for any new company when first manning-up, during plant expansions, or when seeking replacement labor. The new company can *choose* its labor complement after reviewing the qualifications of a number of applicants. Given the scale and variety of the labor on offer, manning-up may be achieved without "poaching" labor from existing companies, thereby minimizing the dangers of competitive bidding leading to wage rises. Over the long term the hiring costs for replacement or new labor may be minimized in the large market not only because the size of the labor pool permits occupational demands to be matched by readily available labor, but also because projected labor requirements can be notified to placement agencies in advance of need so that the dangers of unfilled vacancies are reduced.

The large urban area may offer another type of advantage. Companies which are seeking new locations for plants which are to be manned by executives and key personnel from other regions may search for places with a high quality of physical environment, a range of efficient services, and a lively cultural life. All these assets are more likely to be associated with large urban areas than with small centers. Thus there are a number of substantive reasons why the larger area may offer distinctly lower costs of operation or be attractive on other grounds.

It is more difficult to evaluate the merits of those arguments which suggest that large urban centers also generate external economies providing unique and significant cost advantages to particular sectors of economic activity. Several crucial assumptions are involved. One is that the organizational structure of every advanced economy is continuously being subdivided as each company seeks to gain the economies of product concentration and of large-scale production. The result is that as a percentage of gross manufacturing output, the net value added in each firm tends to fall as companies buy an ever-increasing range and volume of inputs and subcontract larger parts of the production process to specialists. Because of the scale of their activities these specialist producers not only produce at a lower unit cost than nonspecialists but can also hold stocks of semifinished or finished goods for delivery to enterprises with unexpected input requirements. This process, which can be viewed either as a disintegration of the production structure or as the growing interrelatedness of separate production units, is dynamic. Major companies constantly seek to spin off those activities which yield low or risky returns to specialists and are constantly persuaded by these specialists to try new forms of product linkage as innovations or product improvements occur.[13] It is argued that for each production unit a large part of this web of advantageous input-output linkages is associated with a *given location*. Essentially, these local linkages

13. A stimulating discussion of this process is given in J. R. Lasuén, "Growth Poles," *Urban Studies*, Vol. 6, No. 2 (1969).

represent not only the effective capability of suppliers to produce at a given price and to a given quality but the explicit awareness by suppliers of the production techniques and operational constraints of the buying enterprises. Indeed, since effective supply linkages involve not only initial learning costs but also constant reevaluation by the supplier of the opportunities for new types of output, then personal contacts between buyer and seller are the essential means of developing and strengthening the mutually beneficial relationship. Typically, these contacts are thought to be between producers within the manufacturing sector, but it may be that relationships between manufacturers and public utility suppliers for services such as electricity, gas, water, and effluent disposal are increasingly important. Whatever the exact components of these local input-output linkages, it is argued that the declining protection from competition caused by improved transport and communication facilities requires every company to reduce costs by finding a location which guarantees technical external economies of the kind discussed above.

The critical premise in this argument is that these technical externalities develop as a direct function of the growth in the urban community. With large community size comes a variety of demands of sufficient aggregate scale to initiate the development of specialist markets. Size also confers a greater probability that for any given specialist activity there will be a number of producers so that there are real possibilities for price comparison and competition, as well as an absolutely larger volume of spare specialist capacity at any one point in time. It may also be that a large range of specialist producers in close physical proximity to each other encourages rapid diffusions of technological and product innovations.

The cost penalties incurred by any individual company which chooses to locate in a free-standing and small urban area are obvious and inevitable. At best the maintenance of contact with specialist producers in the large centers may be bought at the price of increased transport and communication costs. At worst this suboptimal location may result in imperfect knowledge of the range of specialist suppliers and specialist expertise. Given this view, the decision to locate away from major centers cannot result in a reduction of operating costs.

In summary, the defense of a growth center policy based on relatively large centers is that external economies are increasingly important for the average firm, that close access to a range of specialist suppliers not only minimizes friction and storage costs but also induces mutual product and production innovation, and that such economies are most likely to be maximized in large urban centers. Furthermore it is argued that the external advantages of agglomeration are more significant than any specific advantages which smaller centers may possess such as lower efficiency wages or lower local taxes. If this argument is valid, all new exogenous activities should be encouraged to locate in the relatively large population agglomerations and steered away from small, high-cost urban centers.

A conclusion as inflexible as this is invalid. Certainly for the new company just starting up, or the totally transferred undertaking, great reliance may be placed upon the range of potential local suppliers and subcontractors, since the quality of these local linkages may have an important influence on the performance of the company in the crucial stages of winning new customers or of reestablishing contact with existing buyers. However, in other types of company, local executives may have neither the freedom nor the necessity to establish close links with local suppliers, ancillaries, and subcontractors. There is no one reason for this situation, but the critical element is a scale of operations which permits specialization *within* the organization. For example, companies may operate production "enclaves" which work up semifinished goods produced within other parts of the organization, thereafter distributing the finished item to the local market or back to a central distribution division. In such cases the demands for local goods and services may be very simple. Alternatively, even though the company is structured into distinct divisions, each of which turns out a separate product, decisions relating to input purchasing, subcontracting, finance, insurance, and so on may not be a divisional responsibility. One variant of this is the situation in which the new division is established for a trial period with inputs supplied by long-standing associated companies from their existing locations. If the new venture is a success and the forecast scale of operations materializes, then existing suppliers may, in time, establish a linked production unit in the new location. In any event, whether the company services its local plants itself or through other firms, the freedom of local plant executives to negotiate with local suppliers is strictly restrained by policies which are designed to fulfil the proper functioning of the whole corporation.

There also must be some doubt as to whether a location distant from major complexes or linked suppliers necessarily entails a markedly inferior cost structure. This must depend upon the distributive pricing system used by the suppliers. If the major costs are those of vehicle acquisition and depreciation and of goods assembly and dispatch, and there are insignificant differences in line-haul costs to individual customers, then delivery charges may be standardized over a wide market area. The result is that accessible customers subsidize more distant customers and thereby diminish the incentive for a location in close proximity to the supplier. Furthermore, personal contact with suppliers and subcontractors need not always be won at the price of manufacturing plant location which is highly accessible. Instead, the establishment of an accessible purchasing/subcontracting office may suffice.

The gist of this whole argument is simple. A priori, the relatively large urban areas seem to offer significant cost-reducing, market, and environmental opportunities for the majority of migrant companies and especially for new and transferred companies. It follows that regional redevelopment authorities are most likely to maximize the flow of migrant companies for a given subsidy cost if they concentrate their "advance" subsidies for infrastructural investment in these larger centers. It does not follow that smaller, relatively remote centers will be bypassed by all migrant companies, since they may offer to producers with little need for local inputs and subcontract assistance a relatively low-cost environment based on cheap labor. Thus the redevelopment agencies should be prepared to offer supporting infrastructural assistance to such areas and direct assistance to companies after they have chosen their new location. This point will be discussed in more detail in the following sections.

Long-Term Economic Growth

If the arguments of the preceding section are valid, a growth center approach will *divert* activity from competing prosperous regions at a low subsidy cost.[14] It also follows that existing external economies, which were influential in attracting exogenous capital to the center(s), automatically will be enhanced as the scale of the active labor force is expanded. Consequently the new export industries should begin their operations in highly productive conditions, thereby contributing to their short-run efficiency and the growth of the local economy. Thus the discriminatory injection of public funds should result in an imbalance in the performance of different urban areas within the region, with the growth center(s) acting as the

14. I.e., such a policy will increase the regions' share of national net value added by more than if a growth center policy had not been in operation.

spatial "cutting edge" of a drive towards increased regional output and per capita productivity.

Acceptance of this chain of events, however, does not establish whether the growth center approach will encourage the fastest rate of unsubsidized growth in net regional value added over the long run. Certainly a previous chapter has suggested that the response *within* the growth centers to the enhanced opportunities is likely to be favorable. Several factors were identified as "spurs to economic growth and efficiency," ranging from the massive flow of technical, economic, and financial information generated by indivisible communication assets, such as trade associations, stock exchanges, and business consultants, to "personal interest" investments made by powerful individual and corporate entrepreneurs.[15] However, it is not at all clear how companies, and financial agencies outside of the growth centers, will react to the concentration of activity. Very broadly there are two possibilities. The optimistic interpretation is that areas which are deprived of their fair share of public investment will suffer a relative reduction in the quality of their urban services and consequently a diminution in local external economies. As a consequence, capital which might have been attracted to these areas is invested in the favored growth centers. Labor and capital also migrate out of the declining areas to the growth centers under the lure of higher real earnings. However, even after allowance is made for the costs of intraregional migration and resource transfer, the region can expand its per capita net output and its per capita income. Moreover, although in the secondary areas there is a fall in the level of employment and of activity, skilful adaptive planning can create new economic structures based on new exports to the growth center(s) and to other regions. Hence there is no reason why *per capita incomes* in the secondary areas should not rise at a satisfactory rate.

The alternative and pessimistic theory is that a growth center policy may not maximize the long-run rate of regional growth simply because it is impossible to select the right growth centers and/or because capital deprivation results in a drastic fall in the level of economic activity in the neglected areas.

The first of these criticisms, which can be made against any form of spatial discrimination by governments whether national or state, assumes that even the most proficient agency cannot predict where growth will occur. Thus a series of seemingly unpredictable chance

15. See Chapter 2, pp. 28-32.

factors may make nonsense of the most rational choice and confound intricate technical judgments based on input-output analysis, employment projections, production cost comparisons, and so on. Failure to choose the right centers has a high cost in terms of alternatives foregone, since there is a great danger that massive concentrations of scarce public investment may create excess capacity in particular locations and severe shortages in other locations, with obvious effects on regional costs.

It is also argued that spatial discrimination may spell the death knell of efforts in local development planning and particularly in those communities distant from the main centers. There are two reasons for this. Whatever the exact form of raising the funds for public investment, a growth center policy typically results in growth center inhabitants receiving marginal public benefits in excess of their marginal tax payments. The reverse of this may apply to taxpayers in the neglected centers, who find that the real benefits from paying taxes are reduced. In time the concentration of investment is likely to encourage an increase in the flow of exogenous enterprise and investment to the growth centers, and a reduction in the flow to secondary centers. Neither of these conditions is likely to make the efforts of secondary center redevelopment planners any easier. In particular, the reduction in the inflow of exogenous enterprise may deprive local planners of one of the most powerful mechanisms for transforming the performance of key local production sectors. Thus although the obvious result of an enterprise inflow is an increase in the net incomes of construction workers, plant operatives, owners and workers in tertiary services, and public authorities, the crucial benefit may be that local producers are forced to compete for labor. Normally incoming companies attract the bulk of their labor from other local producers,[16] and some of the unemployed replace the workers who have quit, thereby receiving training in relevant skills. Now if this influx is on a large scale (in terms of either the number of new plants or size of demands by a few new companies in relation to the available supply of labor) then the bidding away of labor may force companies to face the alternatives of liquidation or improved productivity. The migrant inflow therefore may be seen as a crucial lever for inducing economic change.

It may also be claimed that a growth center policy distorts free competition. Concentration of public investment may confer a price

16. See, for example, the article by E. M. Jones, *Manchester School of Economic and Social Studies,* Vol. XXXVI (1968), pp. 149–63.

advantage on growth center producers, some of whom are not as efficient as competitors in the secondary centers. The effect may be to encourage the expansion of the least efficient at the expense of communities less able to accommodate thwarted growth in leading sectors.

The final argument is that even if the discriminatory investment policy does result in a reduced flow of new migrant industries to the neglected areas, the unemployed and underemployed will not migrate in any numbers to the regional growth centers. The case, a familiar one, assumes that low-income workers will consider that their rewards will not be increased by geographical mobility, that the unemployed are ignorant of job opportunities elsewhere, and that many others are tied by previous investments, job experience, or personal commitments. In aggregate, these restraints on movements are assumed to increase exponentially with migratory distance, and since the regional center may be up to a hundred miles away, the flow of migrants may be very small. It follows that the regional centers cannot rely on constant infusions of hinterland workers, so that with massive public investment and inflows of exogenous capital, demand-induced inflation becomes inevitable.

The logical prediction from all these arguments is that a growth center policy will create oases of prosperity within regions which are generally depressed. It follows that remote local areas deprived of their legitimate share of public investment not only are justified in using political pressure to undermine the policy of discrimination, but ought, on grounds of regional growth, to use local subsidies to attract migrant industry. Such a strategy will initiate the maximum revitalization of inefficient local activities and provide the most diverse opportunities for experimentation in local redevelopment. Thus, although the subsidy cost of initiating development may be higher than under a growth center approach, many of these local experiments may succeed over the long run and create far greater opportunities for broad-scale revitalization than would emerge from an approach which freezes the smaller centers into the role of ancillaries of a development process generated elsewhere. Over the long run, then, a policy which scatters investment widely will maximize the difference between the costs and benefits of subsidization and will create far greater personal involvement with the revitalization of declining economies.

The validity of many of these points is suspect. Provided the selection of growth centers is not biased toward green field sites, or small centers with no previous record of fast growth, or centers which are badly located in terms of interstate and intrastate communications networks, the choice, in agricultural or mineral resource hinterlands, is likely to be very limited. Similarly the dangers of thwarting the growth of efficient economic sectors and companies can be exaggerated. Since one objective of the growth center policy is to free human resources from unproductive activities, public support for efficient companies in the growth center may hasten the demise of inefficient producers. A similar point can be made against any argument based on the assumption that all local communities are interested and proficient in devising redevelopment strategies. In some instances the most honest policy is actively to plan for the decline of particular communities which have no real chance of revival, or at most treat them as commuter towns. There is no reason to suppose that the attractions of the chosen centers will be inadequate lures for many hinterland earners. Apart from the security and choice offered by a varied range of job opportunities and the high quality and wide range of urban services, the regional center may be the preferred locale for assimilating the city culture while maintaining personal, and perhaps economic contacts with the original hinterland location. Of course the impetus to intraregional migration would be strengthened if the uncertainty and costs associated with movement were minimized. Cash grants to cover the costs of moving, labor market and housing market information, assistance in selling houses in the original location, subsidized job training—all of these would increase the flow in the right direction.

No equally simple arguments can be used to destroy the argument that concentrated investment will result in the cumulative decline of potentially viable communities located at some distance from the growth centers. Obviously much depends upon the efficiency of communications between the primary and secondary centers, the degree of complementarity of economic activities, the responsiveness of business leaders in the secondary centers to the enlarged business opportunities in the center, and the extent to which the secondary centers can protect and develop their function as tertiary service centers.

Regional policy should not, of necessity, be founded upon the notion that a reduction in the volume of population and of activity in certain areas will automatically imply a reduction in per capita productivity there. Indeed physical investment programs for the growth centers may be paralleled by various subsidies to improve the marketing, production, and distribution of the existing local sectors located in the secondary centers. Secondly, there is an obvious role for investment in human resources development which permits the recipient an element of choice in terms of the ultimate location of skill utilization. Finally, there remains the distinct probability that the flow of exogenous investment to the secondary centers will not cease entirely. There is therefore an obvious case for financial aid to companies which choose such locations, provided, of course, the local subsidies which have lured them there are not exorbitant in comparison to the subsidies in the growth centers.

Planning Advantages

Typically, the growth center will be designated as part of a total planning area which extends far beyond the boundaries of the center itself and includes a variety of urban settlements of varying economic structure, size, and economic performance, each of which may be under the planning aegis of distinct authorities. Given this wider planning horizon, the growth area planners may be in a strong position to delineate viable planning objectives for the whole area from an information base which identifies economic trends, the likely locational impact of public investments, the varying planning objectives of each authority, and the future financial resources available for public planning policies. A priori, it seems reasonable to expect economies of scale in plan formulation, the elimination of blatant inconsistencies in the objectives of different authorities, or, at the very least, enlarged possibilities for mutually beneficial investment schemes.

However, the major advantage is that any planning is hierarchical. Since the growth center itself is seen as the proper environment for major public investment, all other parts of the planned space are ancillary to the economic development process which is to be created within the center. This may permit the formulation of explicit economic activity and population targets for the center, the specification of future investment strategies, and the identification of those areas in which future population and activity growth is likely. After this stage, the necessary planning implications for the economic life of the ancillary centers can be characterized, either in precise resource and locational terms or more generally.

This precise formulation of planning strategies and location of public investments should permit constant reevaluations of the quality and direction of planning.

Conclusions

Five arguments in favor of a growth center strategy for the most distressed regions have been analyzed, and though all of them lack empirical substance, their merits on a priori grounds are clear. Thus any policy which contributes in the long run to a more rapid concentration of a region's population into a few large urban areas is likely to create the conditions for servicing net and replacement demand for social/economic overhead capital at a low per capita cost. Moreover, for a given subsidy cost, spatially concentrated investment is likely to maximize the flow of income to regional earners in the short run; to attract the maximum flow of exogenous enterprise and capital; and to generate a highly productive environment in which an expanded export base can reduce the regional balance-of-payments deficit and provide sufficient job opportunities to restrain the flow out of the region of the economically active. Finally, the quality of short-term regional planning may be improved if the mix and scheduling of public investment over time is given a rigorous spatial dimension.

Despite these apparent advantages, considerable uncertainty clouds various aspects of the policy. Partly these are technical uncertainties. What is the best size for a growth center? How is optimality to be summed from a series of separate conclusions on the best size for minimizing per capita infrastructural costs, for maximizing external economies, for generating the greatest spill-over benefits, and for initiating long-term growth? How can purely economic considerations be matched with social considerations such as the personal satisfactions of growth center residents, which may be at their greatest in relatively small communities?

The simple fact is that we have few guidelines which can help shape our answers to these questions. Our examination of two major studies confirmed that the unit costs of providing social and economic infrastructure were relatively high in small population concentrations, but there was no agreement on where the threshold of sharply rising unit costs occurred in large communities. In one study no marked increase in costs appeared to occur in communities of between 130,000 and 1.1 million, whereas in another study unit costs were seen to rise sharply in communities of over one quarter of a million. When we turn to the optimal size of communities for generating long-term economic growth within the center and for encouraging growth in the hinterlands, there is a complete lack of rigorous research. Certainly, one detailed projection has suggested that counties with a population between 50,000 and 500,000 will have a growth in employment up to 1975 in excess of the national rate of growth,[17] largely because such counties possess a favorable weighting of fast-growth employment activities. This would tend to suggest that population concentrations of this size should be chosen as growth centers. However, this can be adopted only as an approximate working rule, since the same study established that a proportion of counties *in every size range* were likely to grow faster than the national average growth rate. In addition there may be cases where a selection must be made from centers which do not have a favorable growth structure, so that size is likely to be only one of the variables which must be considered. Thus, the essential rule is to avoid concentrating investment in very small, remote centers but to remain alive to opportunities in any other center including those of 1 million population or more.

A growth center policy must also face intricate political and organizational questions. What kind of planning body should direct growth center development strategies? How should the body be selected, elected, or formed? What should be its relationship to county planning, city planning, state planning, and perhaps regional planning? In what ways should Senators and Congressmen in the Federal government be involved in the work of the growth center planning agency? Above all there are uncertainties relating to the impact of a growth area strategy upon areas within the region which have not been chosen for prime investment. The optimistic view is that the growth centers and secondary centers can benefit from concentrated spatial investment. But there are legitimate grounds for assuming that complementary policies for the redevelopment of the economic base of secondary centers are absolutely necessary, if labor and capital underutilization is to be prevented. These political and organizational issues are raised and discussed in Chapter 6.

17. U.S. Department of Commerce, EDA, Apr. 1967.

4

THE EVOLVING FEDERAL ROLE—THE LEGISLATED FRAMEWORK

Previous chapters have dealt with the main directions and components of a Federal regional policy which might maximize national efficiency over the long run. However, only limited reference was made to the socio-economic characteristics and scale of the distressed area problem and no explicit attempt was made to evaluate the political possibilities of pursuing the necessary economic strategies.

As a prelude to a final summary dealing with an *economically reasonable and politically realizable strategy,* the next three chapters seek to discuss the main political constraints within which a Federal regional policy must be formulated and implemented and to analyze the policy options on which administrative choices must be based. The first chapter initiates this discussion by focusing upon the Public Works and Economic Development Act of 1965 and analyzes the nature of the justification for Federal involvement with distressed areas, the policy objectives, the authorized development strategy, and the types of institutional framework thought necessary for the achievement of the policy objectives. In the next chapter, a detailed typology of distressed areas is developed in the context of the legislated criteria of distressed area status. In the chapter which follows, the role of the agency prosecuting regional policy, the Economic Development Administration (EDA), is examined in the light of major policies selected by the agency in the period from the inception of the legislation to the end of the Johnson administration. And finally the involvement of the states with programs of area revitalization is analyzed in the context of the current operations of the agencies chosen by the legislation to coordinate and focus the efforts of the states and the Federal government in multistate planning and programing (the Regional Action Planning Commissions).

Certain Critical Principles in the Legislation

The clear and oft-repeated objective of the act of 1965 was the achievement of an economic, as distinct from a welfare solution to the problems of lagging regions and areas, through a combination of economic planning and program implementation at the local level backed up by technical assistance and broad-gauge subsidization by the Federal government.[1] One justification, which was often mentioned in the congressional debates prior to the passing of the legislation, was that subsidies for distressed areas are ultimately offset by a diminished need for transfer payments and an increment to national output and thereby national taxes which come from putting the unemployed to work. However, the major long-term argument was that the use of Federal subsidies and active Federal participation in subnational revitalization were central components of an overall planned process which would reduce local supply costs, attract private capital, and recreate the local economic base. In consequence the legislation constantly used terms such as "long-term economic development," "diversified and stable economies," "permanent jobs," and a "Federal-local partnership in planning." Thus the claim was that the use of subsidies within a planned framework would not only offset current comparative deficiencies but result in net competitive advantages which would attract new economic activities.

This Federal commitment was based on assumptions which were seen in Chapter 2 to be integral parts of the theory of "planned adjustment." The first assumption was that spatially concentrated personal hardship could not be eradicated by unaided market forces, so that there must be public planning and programing to identify structural weaknesses and to provide the necessary public framework for eliminating these weaknesses. Secondly, it was assumed that the distressed regions and areas themselves must create the necessary institutional structures through which local citizens can express their views on the objectives and strategy of redevelopment and in which ultimate responsibility for the selection of area economic goals and on-site implementation of development strategies can be seen to reside. Thirdly, it was assumed that the task of redevelopment could not rest solely with local agencies, partly because of local deficiencies in redevelopment know-how and financial resources, but also because local competitive subsidization might waste resources on fruitless searches to attract footloose industry. Thus the Federal government in general, and its executive agency

1. This discussion does not examine Federal regional policy prior to 1965. For a very full and interesting account of earlier Federal efforts at area revitalization see Sar A. Levitan, *Federal Aid to Depressed Areas: An Evaluation of the Area Redevelopment Administration* (Johns Hopkins Press, 1964).

the Economic Development Administration in particular, was seen as a catalytic power activated from below. EDA was given legislative authority to encourage, guide, and supplement the redevelopment efforts of local planning agencies, provided these agencies prepared an overall economic development plan and suggested realistic projects which would create a net increase in economic activity, rather than divert activity from other areas. Finally it was assumed that Federal assistance must be sufficiently comprehensive so that the varying needs of each distressed community at different stages of the redevelopment process were covered. Thus the legislative authorization provided support for the establishment and operation of local planning agencies, technical assistance grants to help identify structural weaknesses and opportunities, and public works grants to improve the provision of social and economic overhead, and for business it provided working capital guarantees and low interest loans provided they created employment in designated areas.

The broad objective of Federal involvement follows from these assumptions. Thus the goal was to minimize unemployment in the designated areas of high unemployment and to maximize median incomes in designated areas of low income within the constraints of fixed Federal funding and the strict obligation to avoid sponsoring activities which diminished employment/earnings opportunities elsewhere. Moreover, the act continually stressed that every designated community had the right not only to ask for Federal assistance but to expect that realistic project claims would be approved. Such requests for Federal assistance could emanate from any of three spatial groupings. A representative development agency in any individual designated area could negotiate directly with the Economic Development Administration for project and technical assistance, as could any private company which wanted business loans, provided it already operated or intended to operate within a designated area. Alternatively, a group of two or more redevelopment areas could formally unite with contiguous non-redevelopment areas and be designated as a "development district," provided two conditions were fulfilled. The whole district must contain a sizable urban center in which public investment could be expected to spur the development of the whole district and a representative executive planning agency which had the authority to formulate redevelopment strategies for this district. The formation of such a multicounty grouping required the initial permission and, wherever feasible, the active cooperation of the state or states concerned. Finally, a redevelopment area could be selected to receive public investment as part of an agreed multistate strategy initiated by one of the Regional Action Planning Commissions. These agencies were to be sponsored by the Federal government and contiguous states in the various lagging regions of the country and to have as their objective the preparation of plans and initiation of economic development programs, in which the impact could be reasonably expected to affect some or all of the participant states.

Some Unanswered Questions

Superficially the objectives of the act, and the spatial and institutional strategy which it endorsed, appeared to be clear cut. On closer scrutiny, the legislation was open-ended on several crucial issues. Granted that the overall objective was to aid the development of new export bases in the distressed areas, one critical question is whether this was to be pursued primarily as an economic, a social welfare, or a political objective. For example, did the act imply that the nation's rate of growth, measured in terms of per capita income, would be faster over the long run if subsidies were used to create work in the areas of surplus labor resources than if unbridled expansion were permitted in the regions operating nearer to the margin of capacity? Alternatively, was the act primarily concerned with the social welfare function of redistributing income to people in distressed areas through the provision of subsidized work opportunities rather than transfer incomes, even though this might lower the rate of growth of the national economy? Or was the real objective the maintenance of a multiplicity of political and economic decision centers within the country, on the supposition that this would contribute to the political health of the nation? In turn this raises the question of whether the Federal commitment to aid distressed areas necessarily dictates that the objective to be achieved *in every case* of assistance is the creation of a sufficient number of jobs at a socially acceptable level of wages to employ the currently unemployed and to guarantee work for natural increases in labor supply. If this was not the objective, should the Federal government accept that redevelopment success may require entirely different solutions for different types of area and even include the actual reduction in the size of the active population of some areas? Furthermore, does the act really imply that every designated area must be helped? If this is the case, how should the Federal government ration its scarce resources, if, as can be assumed, requests for project assistance consistently outstrip the available finance and staff provisions? Clearly this is related to the development strategy which the agency follows. Should it, for example, encourage redevelopment areas to submit long-term programs requiring a variety of long-term subsidies commitments and real coordination by different Federal agencies of their investment strategies for such areas, or should each Federal agency attempt to initiate as much across-the-board local activity as its funds will permit? Finally, what type of institutional framework, in the short term and the long term, should the agency seek to construct for the initiation and implementation of its regional policy?

DISTRESSED AREAS AND NATIONAL OBJECTIVES

The whole thrust of the act seems to imply that the permanent solution for the distressed area problem can be initiated by a marginal redistribution of new economic activity from the prosperous areas to the distressed areas and by encouraging indigenous activity within the distressed areas. This argument was based on three premises: first, that subsidies can be used to prevent a cumulative loss of resource advantages during the period of the formulation and implementation of a planned program aimed at permanently raising the level of demand for local labor; secondly, that the underutilized resources in distressed areas provide the supply conditions for a rapid rise in real output in the short run; and thirdly, that over the long term the competitive disadvantages of such areas will be removed by internal action so that private capital will flow in without continuing subsidy inducements. It therefore appears that the policy objectives were primarily economic in that the initial use of subsidies was expected to raise national output over what would have occurred without subsidization. But over the long term, can this policy of spatial redistribution of economic activity be justified on similar grounds? In fact the act did not predict whether national output would be unchanged, higher, or lower than it would have been without direct action at the distressed area level. However, there are several clues in the legislative prose which suggest that the real claim was that the economy would grow at a faster rate, precisely because resources would be used more effectively in the distressed areas.

It is important to note the nature of this claim. The act did not seek to justify Federal involvement with distressed area job creation on the ground that the more prosperous areas have reached, and perhaps gone beyond, an optimal stage of growth. Thus the distressed area problem was viewed as an aberration within a generally efficient economy which can be resolved without impairing the anticipated and necessary growth of the faster growing regions. Furthermore, since the market allocates the bulk of economic activity to optimal locations there is no value in formulating comprehensive Federal policies for the future distribution among the regions of the nation's population and economic activity. Thus the core objective seems to be to revitalize certain decision-making *locations* which would stagnate without national assistance, not only because this is an aid worthy in itself but because it will also strengthen the national economy.

But this still does not dispel the uncertainty of whether the precise goals for each distressed area are necessarily the same. Of course, the easy way out of this dilemma is to assume that each designated community charts its own goals and that Federal responsibility ends when the area in question cannot qualify under any one of the several qualification programs. However, does this mean that for the low income areas, as an example, the goal of maximizing median family incomes must be pursued for a given number of economically active persons? Similarly, with the areas of high unemployment the achievement of a satisfactory level of employment may be dependent upon encouraging outmigration which actually reduces the local labor force. Thus the real question is whether the major objective of the policy, which seems to be a faster growth in national output, is consistent with attempts to redevelop every distressed area within homogeneous constraints.

SPATIAL PRIORITIES

Given the strict limitation on Federal finance and human resources, the need for criteria on which to base fund allocations to distressed areas is obvious. However, the legislation was highly ambiguous on this. There are several allocation procedures which might be used. Priorities could be based on a simple queuing system so that those areas with reasonable projects and an early place in the queue would receive prior treatment. Alternatively, the priority criteria could be much more explicit and could be based on deliberate Federal choice. For example, the seven distinct methods of qualifying for distressed area status could be ranked in order of "importance."[2] On this, however, the act is silent, so that it must be assumed that the legislators were more concerned with including every area with any possible ground for assistance than with efforts to define whether low income problems were more urgently requiring solution than problems of high unemployment.[3]

Another method might be to concentrate upon projects or areas in which the benefits per public dollar (of subsidy) seem to be highest or upon those areas which exhibit the greatest symptoms of distress. Theoretically resolution of this conflict depends upon assumptions about the marginal utility of increased employment/income derived from subsidies under varying socio-economic conditions. If the marginal utility of expected benefits is thought to vary inversely with the initial area level of local employment and/or income, then welfare gains will be maximized by concentrating upon the areas of greatest apparent distress, however this is measured. If, on the other hand, the marginal utility of increased income/employment is thought to be constant at any initial socio-economic level, then welfare gains will be greatest when public investment is focused on those projects (and areas) which maximize the difference between the costs and benefits of investment.[4]

The distinctive feature of the legislation is that particular clauses gave some endorsement to several of these competing methods. As previously noted, the legislation suggested that any designated area can receive assistance, and this might imply a simple queuing strategy of subsidy allocation. However, the act also indicated some form of preferential treatment for the severely distressed areas, insofar as they were to have eligibility for the maximum amounts and widest range of subsidies. Certainly, it was not made explicit whether this preferential treatment was occasioned by the assumption that the benefits per public dollar invested were likely to be highest in the most distressed areas or, alternatively, that as the severely distressed areas had the most inadequate public resources to achieve redevel-

2. See Chapter 5.

3. Of course, insofar as the low income areas and those with net loss of population qualified on statistics drawn from the 1960 census, they therefore cannot be removed from qualification. This is not the case with unemployment areas, so that in one sense Congress accepted the idea that the low income/migration problems were more urgent.

4. Concentration upon projects which bring the highest return can be accomplished by a ranking of projects from any group of areas. Alternatively the areas in which investment is likely to yield the greatest returns may be selected first and then project requests from these areas ranked.

opment then Federal assistance must automatically be larger. Whichever view is taken it is clear that Congress wanted to give especial encouragement to the "worst areas."

This preferential treatment, however, was not coupled with any directive to concentrate upon the most distressed areas to the exclusion of the more prosperous areas. Indeed, a similar kind of preference seems to have been shown for areas in which the returns for investment could be expected to be high, and especially those areas which were part of a regional development district grouping. In this the explicit objective was to encourage the development of medium-sized urban centers which would draw population and activity from designated counties with low prospects of development. The act not only encouraged the formation of such planning areas through a program of fixed annual grant allocations for investment in development centers, whether these were in designated areas or not, and through grants for the operation of planning agencies, but also provided a "bonus" level of public works grants for individual designated areas which were willing to form a development district. Once again, although the legislation constrained the agency from investing in development districts to the exclusion of other areas, it showed clear preferences for the development of a program which was expected to generate considerable economic benefits.

THE RELATIONSHIP BETWEEN THE PLANNING INSTITUTIONS

A final uncertainty is the nature of the long-term relationship between the different types of planning institution envisaged by the act. It is not at all clear whether contacts between the Federal and substate level, that is, individual counties and development districts, were expected to be developed as a permanent part of the area redevelopment process or whether the Regional Commissions were to become the superior planning agencies which gave formal approval to the actions of these smaller units of government. Moreover, the role of EDA in coordinating the activities of all other Federal agencies toward the achievement of a common Federal redevelopment strategy for given areas was not made clear.

Consensus and Legislation Form

The complexities in the structure of the act of 1965 were not occasioned by overt conflict between political groups over the legitimacy and direction of Federal regional policy but rather by the comprehensiveness of the legislation in safeguarding at least part of the interests of *all* the potentially opposed groupings. Thus the act represented a special kind of consensus politics—a consensus based less on a compromise of long-standing conflicts between opponents than on the simple aggregation of a series of contrasting developmental strategies. The genesis of this particular legislative form can be traced to the political environment in which the bill was drafted.

With the Area Redevelopment Act due to expire in 1965, the administration introduced a resolution into the House[5] in early 1963, seeking increases in authorizations for business loans, public works loans and grants, and technical assistance. In May of that year the House Banking and Currency Committee approved the measure and reported it out by a large majority.[6] However, in June the bill was narrowly defeated on the floor of the House by a coalition of southern Democrats and Republicans who alleged that inefficient program administration had resulted not only in considerable waste in particular projects but also in a failure to use appropriated funds fully so that the basic conditions of the distressed areas remained unchanged.[7]

The nature of these allegations is important. Very few representatives openly recognized that the delays in using up appropriated funds and the alleged inexpertise on project appraisal may have been due in part to the legislated redevelopment strategy which forced the Area Redevelopment Administration to work on a county-by-county basis and to implement renewal programs through part-time, voluntary, and largely nonprofessional local groups.[8] Furthermore, public efforts to shift new economic activity from those areas which would have been selected by the creators of the activity are likely to succeed only when the economy is expanding rapidly and approaching the margin of capacity. That such conditions did not exist in the initial years of the Area Redevelopment Act suggests that even the most efficient administration of the act would have

5. H.R. 4996.

6. 17 to 6.

7. One interpretation of this reverse for the administration was that it represented southern reaction to President Kennedy's strong stand on civil rights. While this may have been one element aiding the coalescence of opposing forces, a large number of northern Republicans, who had voted for the Area Redevelopment Act of 1961, changed their vote in 1963.

8. Robert Wrigley, Jr., *Journal of the American Institute of Planners,* Nov. 1964.

brought highly limited benefits to the distressed areas. But whatever the relevance or validity of the criticism of the Area Redevelopment Administration, none of the critics expressed a desire for a fundamental reappraisal of the objectives of the act or even for a review of the basic spatial and organizational framework for the allocation of Federal subsidies. This is hardly surprising. Support for the notion that every community can plan for its own prosperity has always been strong in the Congress and especially in the House. As such it is a political philosophy which restricts the role of Federal and state bureaucrats in the determination of project suitability and creates the environment in which congressmen can play an active part in the formation and operation of local development committees, in the delimitation of redevelopment goals, and in the sponsoring of Federal involvement with renewal strategies. Success in winning Federal approval for particular projects may improve the power base of the congressman. Failure may have precisely the opposite effect. Thus most congressmen are understandably less concerned with considerations of optimality in the spatial distribution of economic activity than with asserting the merits of maintaining the existing centers of decision making; less with establishing priorities for investment as between areas than with activity to ensure that the flow of Federal funds to distressed areas is not interrupted by bureaucratic inefficiency.

In June 1963, a slightly modified version of the House bill was passed in the Senate.[9] The House Banking and Currency Committee acted promptly on the amendments, reporting them out in August, but the Rules Committee refused to grant a rule until twelve months later. But late September 1964 it became clear that the House Democratic leadership did not wish to bring S. 1163 to the vote, since an unfavorable decision was likely again.

Throughout the latter stages of these developments the president adopted a passive role, though it is clear that the efforts of the then Secretary of Commerce to impose a reduction in the staff of ARA were vetoed at the White House level. The real point is that the president and his advisers clearly gave top priority to the legislation for the revitalization of the Appalachian region and did not regard the revision of the Area Redevelopment Act as an urgent matter. Indeed it was not until very late in the presidential cam-

9. S. 1163. The skilful maneuvering of Senator Douglas, who argued that the ARA could not be judged after such a short period in operation, was largely responsible for the successful passage of the act.

paign that the president committed his future administration to seeking congressional support for the extension and strengthening of the Area Redevelopment Administration.[10]

When serious discussion on the future of ARA did take place it was recognized that wide-ranging changes in top-level personnel and a new emphasis on project efficiency would be necessary to regain House support, but that neither of these improvements could answer the fundamental problem of how to concentrate Federal investment in areas with a real potential for economic growth. The solution adopted was the development district approach, which was seen as the context for area-wide planning, under the guidance of professional planning staff and with the long-term goal of investment, population, and activity concentration within the development center.

With the program for Regional Commissions the impetus to action came from neither the House nor the administration. Its genesis can be traced to two groups within the Senate. The majority of senators from states with areas which lagged economically behind the nation as a whole desired to obtain the same opportunities for Federal funds as had been promised in separate legislation for Appalachia. A smaller grouping was convinced that not even the Appalachian legislation was likely to foster the creation of multistate institutions which would have a real coordinating influence on the actions of individual states and Federal agencies. Both groups were determined to have legislation which would guarantee some economic planning at the regional level.

The administration was loath to duplicate the Appalachian legislation in many regions of the country, partly because Appalachia contained concentrations of deep distress and a fairly distinct regional entity, but largely on the grounds of minimizing Federal outlay for planning at a time when a new program aimed at the multicounty level was about to be initiated. However, in order to guarantee a trouble-free passage for the Appalachian legislation, the administration did promise to extend Appalachian-type assistance to other regions of the country.

Separate legislation, which reflected the varying strength of the interest in regional planning, was referred to the Public Works Committee by the Senate (S. 812) and by the administration (S. 1648). Whereas the administration bill authorized the Secretary of Commerce to "invite and encourage" the states to establish regional

10. This commitment was made in a speech at Detroit Airport on Oct. 30, 1964.

action planning commissions, S. 812 authorized the president to establish commissions in any region which had been designated a "development region." For a region to be so designated all that was required was the approval of the governors of two or more contiguous states and a qualification based on very general socio-economic data which showed that the region had lagged behind the nation. The administration bill did not mention development regions, but it did require that any regional grouping should include two or more designated or future development districts. There were also differences in the methods of funding the commissions' administrative expenses and the technical and planning work likely to be carried out by the commissions. In essence the administration wanted to put a limit of $15 million on all expenditures by the commissions, to give powers to the Secretary of Commerce to have Federal aid for planning and technical assistance expenditures returned, and to limit Federal expenditure on the commissions' administrative expenses to the first two years, with an absolute Federal share limit of 50 percent thereafter. S. 1648 designated $15 million for technical and planning assistance alone and "such amounts as may be necessary" for administrative expenses. Furthermore, the bills differed in the extent to which state and Federal membership of the commissions was detailed, the administration bill being vague on membership content and the powers of the members.

Senator Edward Kennedy (Massachusetts) summarized the basic differences between the sponsors when he said:

> "... many of us were seriously concerned during the Appalachia debate that other areas of the country, also lagging in economic growth, should have the opportunity to develop regional programs similar to Appalachia.
>
> "In fact, we were prepared to introduce our own regional amendments to the Appalachian legislation, but held off on this after assurance from the administration that we would be covered in a forthcoming national development bill. . . .
>
> "The administration has wisely chosen to extend the Appalachian approach of regional programming to other problem areas. In my opinion, this is one of the most significant new steps in economic development legislation to come forward in many years.
>
> "I would suggest that the committee seriously consider integrating the language of S. 812 into the administration bill. . . .
>
> "I feel that this language would provide more effective guidelines and greater flexibility in the selection of regional commis-

sions, a stronger incentive for such commissions to come to the administration and to Congress with a development program as soon as possible, and a more specific description of the membership of the regional commissions and of the power of the Federal member in approving commission plans. . . ."

The Public Works Committee adopted the Senate bill almost in toto, and with some minor revisions S. 1648 was passed by the Senate and the House in August 1965.

Distressed Area Designation in the United States

As a result of the precise form of the 1965 legislation the U.S. approach to designation is both complex and comprehensive. This complexity does not result from a particularly unusual selection or mix of distress criteria. Indeed only three basic criteria, unemployment, median family income and net population loss plus median family income, are relevant. Yet the critical fact is that variants of all these basic criteria are used to generate six quite distinct programs of distress qualification. To this is added one "political" program which designates one area in every state not having any areas eligible under a "genuine" distress program. Three programs deal directly with the *relative* phenomenon of above-average unemployment, and in three quite distinct ways. One program is directed at "hard-core" areas which have had persistently and significantly higher than national average levels of unemployment in preceding years.[11] Another program deals with those areas of substantial but recent unemployment which had an annual average rate of 6 percent unemployment or more in the previous calendar year. And the third seeks to identify emerging conditions of high unemployment in areas where "the sudden loss or curtailment of a major source of employment has recently caused, or threatens to cause, an unemployment rate exceeding the national average by fifty percent."[12]

In all these programs distress is a relative term, even though each program applies to a different duration and severity of area malaise. However, in the low income, population loss programs entirely different factors are taken into account. In particular, absolute

11. Unemployment of 6 percent or more in the most recently available calendar year and 50 percent above the national average for three out of the four preceding years, or 75 percent above for two of the three preceding years, or 100 percent above for one of the two preceding years.

12. *Highlights of the Public Works and Economic Development Act of 1965*, House Committee on Public Works, 89th Cong., 1st sess., Sept. 1965, p. 5.

point-of-time figures are used to identify distress. The low income program embraces areas in which the 1959 figure of median family income was 40 percent or less of the national median income, and the low income, population loss group covers those areas with a local median family income 50 percent or less of the national average and a net loss of population between 1950 and 1960 of 25 percent or more. The qualification of Indian reservation areas uses both current and historical income and unemployment data to concentrate on those areas where distress is most severe.

These distress indicators can be applied on any one of three levels of geographic space. Apart from very small areas with less than 1,500 population or insufficient scale to be a "labor area," and large counties and municipalities with more than 250,000 population, all of which are excluded, any individual area can qualify as a "redevelopment area" on its own. It may, alternatively, be linked with other counties in a "development district," which has to contain at least two counties and a center or equivalent redevelopment area which is large enough and has the economic potential to stimulate growth throughout the whole district. The final form of distressed area is the "economic development region," a grouping of contiguous states with geographical, cultural, historical, and economic ties which have lagged behind the nation's economic development.[13] Thus the program includes three of the four types of distressed area groupings originally discussed, that is, the isolated distressed unit, the multicounty growth area, and the development region.

U.S. Distressed Areas in Context

This extensive method of qualification conditioned EDA activity in certain obvious ways. Thus the agency had to be responsive to the needs of an enormous number of qualified areas, the bulk of which had an extremely small labor force and population. Furthermore, although there were and are especially heavy concentrations of qualified areas in particular regions of the country, extensive qualification criteria meant that action had to be taken in every region of the country. But the most significant implication was that such criteria embraced distressed areas with markedly different development problems—a point we shall return to again and again.

As of June 30, 1966, the first date when a detailed review of area eligibility coincided with the approximate attainment of national

13. Titles IV and V, Public Works and Economic Development Act of 1965.

full employment,[14] more than one out of every four counties in the nation[15] was able to apply for Federal assistance. However, these qualified counties did not contain one-quarter of the nation's population. In 1960 their population was 29 million (just under one-sixth of the total), so it is obvious that the average qualified area had a relatively small population. In fact, 70 percent of these areas had a 1960 population of under 25,000 and nearly 90 percent, under 50,000. Of course it is true, as the 1960 census showed, that the median population for all units, distressed and nondistressed alike, was just under 20,000. But this does not blur the fact that the concentration of distress among the counties in the population size range between 5,000 and 25,000 was greater than could be expected from a normal frequency distribution.[16] (See Table 4.)

In a later section we will show that the different types of distressed area are associated with different parts of the country, but taken as a whole the problem areas not only were found in most of the mainland states,[17] and in every one of eight regions,[18] but were

Table 4. EDA Qualified Areas and All Counties by 1960 Population Size

Population size of county	All counties		EDA qualified areas		% of all counties qualified
	Number	Percentage	Number	Percentage	
Under 1,000	20	0.6	0	0	0
1,000 – 4,999	274	8.7	75	8.5	27.4
5,000 – 9,999	561	17.9	187	21.1	33.3
10,000 – 24,999	1,094	34.9	372	42.0	34.0
25,000 – 49,999	589	18.8	160	18.1	27.2
50,000 – 99,999	293	9.3	51	5.8	17.4
100,000 – 249,999	177	5.6	25	2.8	14.1
250,000 – 499,999	61	1.9	10	1.1	16.4
500,000 and above	65	2.1	5	0.6	7.7
Total	3,134	99.8	885	100.0	28.3

14. At this date the rate was 3.9 percent. 15. 885 out of 3,134.

16. U.S. Department of Commerce, Bureau of the Census, 1960 Census of Population, Vol. I: *Characteristics of the Population,* U.S. Summary, p. xxvii, Table J; and Economic Development Administration, "Maximum Grant Rates," July 1, 1966, p. xi. The EDA figures exclude 55 Indian reservations.

17. Of the 48 mainland and contiguous states, 38 had at least one area qualified on "genuine" distress grounds.

18. The eight regions referred to are the Northeast, Middle Atlantic, Great Lakes, Plains, Southeast, Southwest, Mountain, and Far West. See Appendix to this chapter for the state groupings for each region.

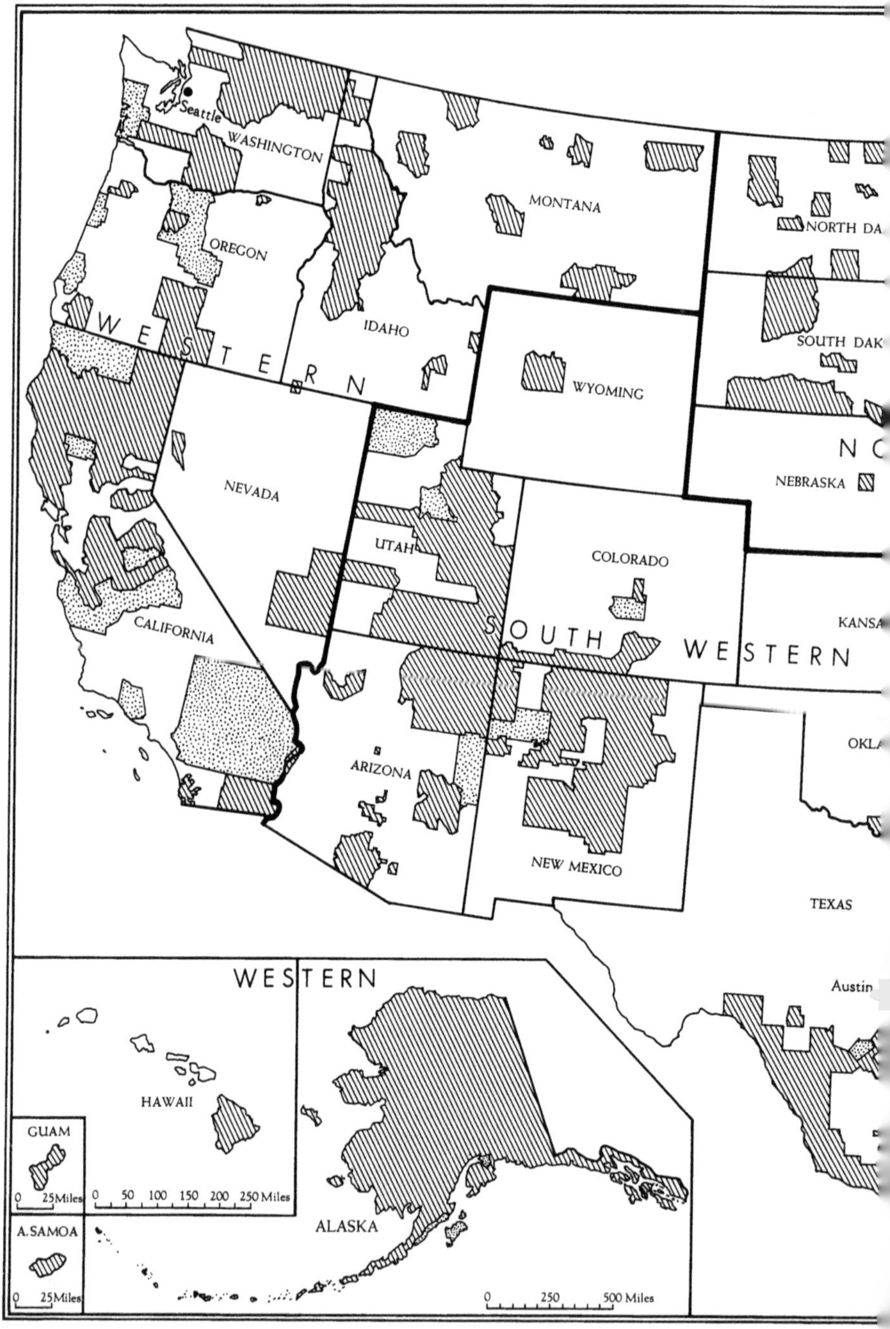

MAP 1. AREAS QUALIFIED BY THE ECONOMIC

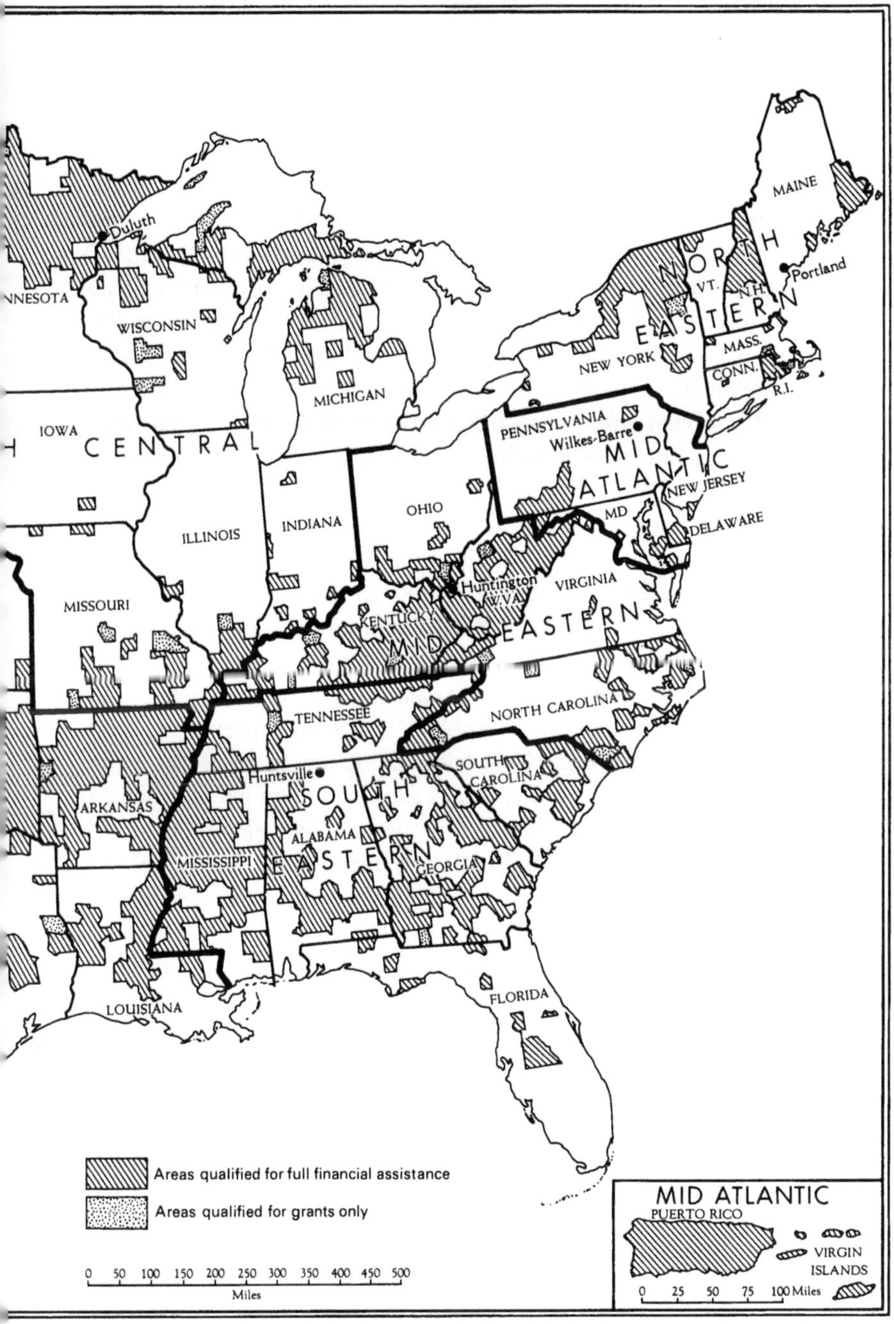

DEVELOPMENT ADMINISTRATION FOR ASSISTANCE

especially concentrated in five regions or subregions of the country, that is, the Appalachian mountain states, the Southeast, the Upper Great Lakes, the Southwest, and even the prosperous Far West, particularly Washington state and California (Map 1, pp. 74-75).

Now we have already noted that area qualification covers two basic types of structural malaise, one characterized by high unemployment, the other by low incomes, and that these basic types were divided into seven separate programs of qualification. Table 5 shows very clearly that there were almost twice as many areas qualified on grounds of heavy unemployment as there were on a low income basis, with the most frequent problem that of persistent unemployment. However, the crucial point is that the two major types of problem area were differentiated not only by economic characteristics but by geographical incidence as well. The unemployment set, in which a limited sector of the working population usually is affected, is volatile in the sense that some areas are constantly being removed from qualification as employment conditions improve while others are being added as conditions deteriorate. For example, in the first year of EDA's operation approximately one-sixth of the qualified unemployment areas were removed from the list because of an improvement in the level of employment.[19] In contrast the low income area problem is not only pervasive, with all or most of the local basic industries having low levels of net value added, but also persistent. Typically, such communities are able to transform the structure of their economic base only slowly, if at all. Of the 60 counties with a median family income one-half or less of the national median in 1950, more than 70 percent continued in this unfavorable relative position at the 1960 census.

Moreover, the amount of geographical overlap between these two types of area was extremely limited. Of the 612 areas qualifying on unemployment grounds, only 50 had median family incomes which were low enough for them to qualify on income grounds as well,[20] a proportion which can be accounted for by random factors.[21] An overlap, indeed, occurred in only two areas of the country, the state of Kentucky and parts of Arkansas and Oklahoma. The unemployment areas were mainly concentrated in the Great Lakes, the

19. EDA, *1st Annual Report*, 1965/66, p. 10.
20. EDA, *1st Annual Report*, p. 10, Table A.
21. See U.S. Department of Commerce, *Area-Welfare-Eligibility for Development Assistance*, Vol. I (1964).

Southeast, the Southwest, and the West,[22] whereas the low income problem was confined almost entirely to the southeastern states.[23] Thus it seems reasonable to pursue our investigation by separating the analysis of the two basic types of problem area.

Table 5. EDA Areas by Program Qualification, June 30, 1966

Qualification criteria	Number of areas[a]	% of all areas	Population (1960)
Unemployment:			
Substantial	153	16.3	8,413,990
Persistent	458	48.7	12,964,277
Sudden rise	1	0.1	13,753
Subtotal	612	65.1	21,392,020
Low income/migration:			
Low income	163	17.3	4,662,413
Population loss	98	10.4	1,573,836
Indian reservations	55	5.9	264,740
Subtotal	316	33.6	6,500,989
State areas	12	1.3	1,275,311
Grand total	940	100.0	29,168,320

[a] Includes 55 Indian reservation areas excluded from the figures in Table 4.

The Socio-Economic Characteristics of High Unemployment Areas

In a recent paper, Chinitz has provided a most useful typology of distressed areas in 1965. His classification of areas of high unemployment has four categories. Table 6 shows the numerical importance of each of these different types of area and indicates some of their socio-economic characteristics.[24] Obviously, the numerically most important areas were the not-so-poor rural areas, though such areas contained only a small proportion of the total distressed population.

Broadly speaking, the high income, fast growth areas were concentrated in the West, and particularly in California. The older in-

22. Concentration in this context means numerical incidence, but it is obvious that such a measure is not weighted by labor force share.

23. For example, in February 1967, 299 areas in the mainland and contiguous states were qualified on grounds of low income or low income and population loss. Of these, 245 (82 percent) were in the Southeastern region.

24. B. Chinitz, "The Regional Problem in the U.S.A.," in Robinson, *Backward Areas in Advanced Countries,* Table One, p. 53. Note that Chinitz's classification relates to areas eligible in 1965 and therefore a very large number of areas are included.

dustrial areas, on the other hand, principally were associated with the Middle Atlantic and Northeastern states. The not-so-poor rural areas were widely dispersed but with marked concentrations in the

Table 6. Socio-Economic Characteristics of High Unemployment Distressed Areas

	High income/ Fast growth	Old industrial	Not-so-poor rural/ Population loss	Not-so-poor rural/ Population gain	Total
Number of areas	38	37	492	105	672
Population (× 1,000)	11,416	17,073	12,548	3,911	44,948
Mean population (× 1,000)	300	461	26	37	
Mean unemployment (%)	6.85	7.6	8.58	7.64	
Mean 1959 median family income ($)	6,014	5,602	4,437	5,037	
Mean % population change 1950–60	+44.7	+7.16	−1.12	+44.8	
Mean population per square mile	282	3,728	68	131	
Mean % nonwhite population	8.6	7.2	7.8	11.9	
Mean median school years completed, pop. over 25	10.9	10.4	9.6	10.2	
Mean % labor force employed in manufacturing	25.0	30.0	19.0	8.8	
Mean % labor force employed in agriculture	4.8	3.7	12.8	9.3	
% population participating in labor force	36.4	38.0	34.8	33.8	

Far West, the Upper Great Lakes, northern New England, and the northern counties of the Appalachians (Map 2, facing this page).

HIGH INCOME, FAST GROWTH

In some respects the inclusion of these areas as distressed is surprising since nearly all the socio-economic indicators suggest buoy-

MAP 2. AREAS QUALIFIED BY THE ECONOMIC DEVELOPMENT ADMINISTRATION AS DISTRESSED, 1965

(From a paper by Benjamin Chinitz in E. A. G. Robinson, ed., *Backward Areas in Advanced Countries: Proceedings of a Conference Held by the International Economics Association at Varenna* [New York: St. Martin's Press, 1969].)

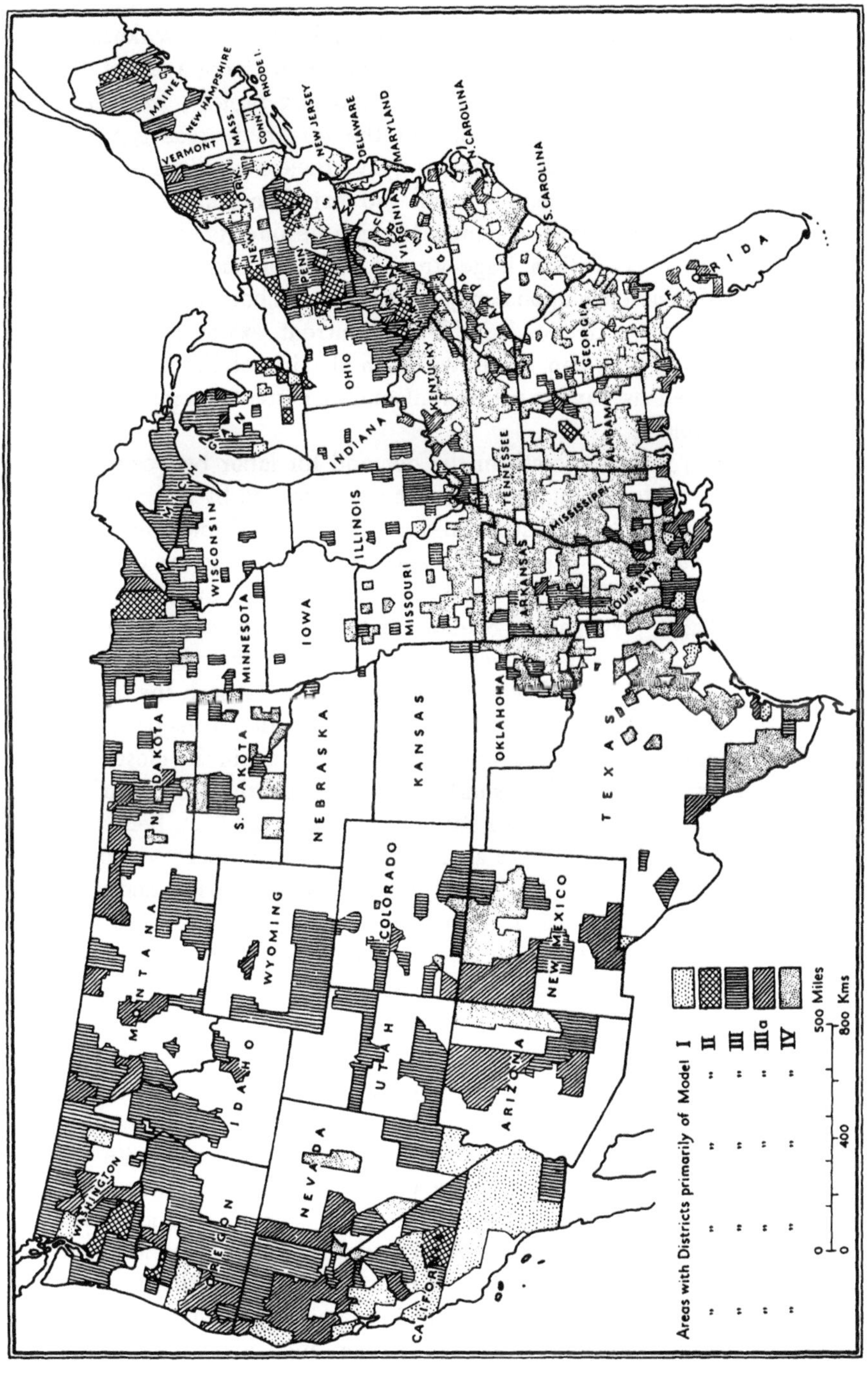
Areas with Districts primarily of Model I
" " " " " II
" " " " " III
" " " " " IIIa
" " " " " IV
0 400 500 Miles
0 800 Kms
MAINE
NEW HAMPSHIRE
VERMONT
MASS.
CONN.
RHODE I.
NEW YORK
NEW JERSEY
DELAWARE
MARYLAND
PENN.
VIRGINIA
N. CAROLINA
S. CAROLINA
FLORIDA
GEORGIA
ALABAMA
MISSISSIPPI
LOUISIANA
TENNESSEE
KENTUCKY
OHIO
INDIANA
ILLINOIS
MICHIGAN
WISCONSIN
MINNESOTA
IOWA
MISSOURI
ARKANSAS
OKLAHOMA
TEXAS
KANSAS
NEBRASKA
N. DAKOTA
S. DAKOTA
MONTANA
WYOMING
COLORADO
NEW MEXICO
ARIZONA
UTAH
NEVADA
IDAHO
OREGON
WASHINGTON
CALIFORNIA

ancy and growth rather than stagnation and structural weaknesses. In most, median incomes were well above the national average, the population was highly urbanized and growing rapidly, from both natural increase and migration, and the qualified areas typically were located in fast growing regions. Nevertheless a serious problem of persistently high unemployment did exist. In some cases this was caused by a growth in demand which despite its scale was insufficient to cope with a huge growth in labor supply. In other areas the dependence upon a major sector led to short-run difficulties when this sector declined suddenly. Current examples of both these types of area can be drawn from the list of SMSA's which were qualified as having substantial and persistent unemployment in 1966 (Table 7).

San José's rate of growth in the demand for labor between 1962 and 1966 was staggering, but this was accompanied by a very rapid rise in labor supply so that an initially high rate of unemployment was not reduced to any extent. An entirely different picture emerges in Kenosha, San Diego, Fresno, San Bernardino, and Stockton. Kenosha had a very marked commitment to a narrow sector of manufacturing (automobile assembly), so that with the closure of a major plant there were few expanding sectors to absorb the unemployed. In San Diego, the local naval base provided a major source of direct and indirect employment, so that here again the closure of a major sector had marked short-term effects on the labor market. With the remaining areas, the structural "shock" has been persistent rather than sudden. All these areas have greater numbers employed in agriculture (and food processing) than in manufacturing, and with the continuing substitution of capital for labor in agriculture and continued growth in the local labor supply, disequilibrium has been persistent.

OLD, INDUSTRIAL

The second type of area, the rich old manufacturing or mining area, is the prototype most commonly associated with European regional programs. Here the causes are well known. Despite its highly urbanized nature, national average incomes, and slow growth or only modest decline in population, its economic structure is dominated by industries which developed during the late nineteenth or early twentieth century. With a national decline in the demand for the products of these sectors, and in some cases the outmigration of key plants to lower cost locations, these areas are left with old

Table 7. High Income, Fast Growth SMSA's with Substantial and Persistent Unemployment in 1966

SMSA	Population				Growth characteristics: % labor force increase, 1962–66 [a]		Economic base: % employed in 1964 in –				1959 median family income ($)
	Rank	% population increase, 1950–60									
		Total	Net migration	Natural inc.	Supply	Demand	Agric.	Manuf.	Selected services	Local govt.	
San José, Calif.	42	121.1	93.3	27.8	29.6	31.5	12.7	29.4	21.7	6.8	7,417
Kenosha, Wis.	198	33.7	15.1	18.6	−6.9	−15.0	9.9	52.4	13.5	5.3	6,916
San Diego, Calif.	23	85.5	58.5	27.0	5.6	8.8	16.5	17.9	25.3	9.0	6,545
Fresno, Calif.	70	32.3	11.1	21.3	15.4	17.9	33.0	11.4	23.9	7.9	5,634
San Bernardino, Calif.	31	79.3	55.9	23.4	20.9	21.6	22.2	13.4	24.3	8.6	5,890
Stockton, Calif.	109	24.5	8.9	15.6	8.5	11.0	23.4	13.1	22.7	8.7	5,889
All SMSA's (224)		27.6	9.8	17.8 [b]	10.2 [c]	12.4 [c]	12.2	27.0	19.3 [c]	n.a.	5,660

n.a. = not available.

[a] Labor supply equals work force; labor demand equals numbers in employment.

[b] 202 SMSA's (excluding those in New England).

[c] Data for 150 major labor markets from U.S. Department of Labor, Manpower Administration, *Area Trends in Employment and Unemployment* (monthly).

plants, an inadequate base of newer industries, narrowly skilled labor, and social and economic infrastructure which is technologically outdated and physically worn out.

Once again, we can discern variants in this basic type of problem area by looking at SMSA's which had heavy unemployment in 1966. The first type of area acts as a service center for a large hinterland so that some of the decline in manufacturing or mining can be absorbed as the tertiary sector increases its share of local net value added (Variant A). In the other areas (Variant B), the proximity to major metropolitan centers may effectively preclude the development of many service activities and compound the problems of structural adjustment.

It is noticeable that in nearly every area the aggregate growth in the demand for labor was in excess of the growth in supply between 1962 and 1966. However, this favorable trend was not strong enough to reduce unemployment to a satisfactory level. (Table 8.)

NOT-SO-POOR RURAL AREAS

Such areas as this, numerically the most important set within the high unemployment areas group, have quite distinctive socioeconomic characteristics. Agriculture, forestry, and mining are major sectors which offer declining opportunities for employment and in which manufacturing is relatively limited. Median incomes are high, but the population is declining or growing only slowly.

One variant of this basic type is the relatively high income rural area with a fast growth in population. Many factors may explain this phenomenon. Rurally located military plants, administrative centers, and new mineral mining activities may account for this situation in some areas, but the major type is the area with highly productive timber and food processing activities in which labor productivity is growing fast. However, this may be caused by capital/labor substitution so that a large section of the labor force is continuously unemployed.

UNEMPLOYMENT AND LABOR MARKET SCALE

Up to this point, distressed areas with high unemployment have been classified in terms of their economic base, their location, and their growth characteristics. But another important way of viewing the distressed area problem is in terms of labor market scale. An analysis of distress in large, smaller, and smallest labor markets is

Table 8. Old Manufacturing/Mining SMSA's with Substantial and Persistent Unemployment in 1966

	SMSA	Population				Growth characteristics: % labor force increase, 1962–66		Economic base: % employed in 1964 in –				1959 median family income ($)
		Rank	% population increase, 1950–60									
			Total	Net migration	Natural increase	Supply	Demand	Agric.	Manuf.	Services[a]	Local govt.	
Variant A	Charleston, W.Va.	107	5.5	−14.0	19.6	4.7	7.7	10.8	24.9	20.0	5.5	5,862
	Huntington, W.Va.	105	3.7	−11.5	15.2	7.2	14.9	14.9	29.0	21.4	6.2	4,977
	Lowell, Mass.	145	n.a.	n.a.	n.a.	7.9	8.7	4.8	41.8	22.3	6.7	6,158
	Wheeling, W.Va.	132	−3.0	−12.7	9.7	2.9	11.4	16.0	26.4	21.0	5.4	5,162
Variant B	Altoona, Pa.	168	−1.6	−11.5	9.9	3.6	7.5	12.6	23.9	18.8	5.5	5,141
	Fall River, Mass.	167	n.a.	n.a.	n.a.	−3.9	−2.6	14.4	43.9	17.0	5.5	5,204
	Lawrence, Mass.	129	n.a.	n.a.	n.a.	2.1	0.0	9.5	51.1	18.6	5.2	6,051
	New Bedford, Mass.	162	n.a.	n.a.	n.a.	4.1	5.0	13.8	44.4	17.7	5.1	5,217
	Scranton, Pa.	112	−8.9	−15.2	6.3	−0.3	5.9	12.6	37.2	19.6	5.0	4,896
	Wilkes-Barre, Pa.	72	−11.5	−17.6	6.1	2.2	7.2	13.0	36.2	17.5	4.5	4,772
	All SMSA's (224)		27.6	9.8	17.8	10.2	12.4	12.2	27.0	19.3	n.a.	5,660

Note: See explanatory notes to Table 7.

n.a. = not available.

[a] Selected services.

highly revealing.[25] The crucial point is that in terms of unemployment many of the small markets have tended to fare very badly despite the growth in the economy, whereas the major labor markets have prospered. We can show this in two quite distinct ways. The first concentrates on areas of substantial labor surplus at several dates within the period 1957–66 and shows not only that the average rate of unemployment has tended to remain at a very high level in the smaller labor markets but that their share in the total number of unemployed has tended to increase. On both scores, that is, the rate and share of total national unemployment, the major markets have improved their position markedly. (Table 9.) Of course it is possible that the major areas' declining share of national unemployment can be explained by a declining share of total employment. However, this does not appear to have been the case. For example, in February 1967, 17 major labor markets and 71 smaller labor

Table 9. Areas of Substantial Labor Surplus, 1957–66

Date	National unemployment rate (%)	Smaller areas		Major areas	
		Unemployment rate	No. of unemployed as % of total national unemployed	Unemployment rate	No. of unemployed as % of total national unemployed
March 1957	4.3	9.0	5.8	9.6	5.4
December 1965	4.1	n.a.	4.9	n.a.	n.a.
January 1966	4.0	n.a.	n.a.	n.a.	4.4
May 1966	4.0	n.a.	n.a.	7.2	2.1
June 1966	4.0	8.0	5.3	n.a.	n.a.
December 1966	3.8	9.3	6.2	n.a.	2.1

Source: Area Trends in Employment and Unemployment, several issues.
n.a. = not available.

markets were classified as having substantial, or substantial and persistent, unemployment by the Department of Labor. Excess unem-

25. The classification into large, smaller, and smallest labor markets is taken from Department of Labor definitions. A major labor area usually has at least one central city of 50,000 population or more, and normally boundaries which coincide with those of SMSA's; smaller areas must have an estimated work force of at least 15,000, with at least 8,000 in employment outside of agriculture. Very small areas fall below these limits but have a population of at least 1,500.

ployment[26] in these areas amounted to approximately 54,000. The total work force in the major markets was slightly less than the total work force in the smaller markets. However, for every (excess) unemployed person in the major market there were 2½ unemployed in the smaller markets. (Table 10.) The really high rates of unemployment were not found in the major labor markets. Indeed, only one major area had a rate in excess of 8 percent. Certainly, the distribution in both the smaller and smallest categories was also heavily

Table 10. Distribution of Excess Unemployed in Qualified Areas by Labor Market Size, February 1967

Labor market type	Number of areas	Est. percentage of total labor force	Est. percentage of excess unemployed
Major labor areas	17	46	36
Smaller labor areas	71	54	64
Total	88	100	100

skewed towards the qualifying threshold rate, but both of these size groups had a large proportion of areas suffering from very heavy unemployment. Almost 40 percent of the smaller labor markets had rates of 8 percent or more, and in one-seventh of these areas, unemployment was at 10 percent or above. With the smallest areas, nearly 50 percent had rates of 8 percent or more, and in no fewer than one out of four areas, 10 percent or more of the labor force was out of work. (Table 11.)

There is also evidence that the majority of the smallest areas with persistently heavy unemployment have not benefited greatly from the growth in the economy. For example, in August of 1963, when the national rate of unemployment was 5.5 percent, 528 mainland areas were classified as having substantial and persistent unemployment. Almost four years later, in February 1967, when the rate of national unemployment had fallen to 3.7 percent, 263 of these areas remained in this persistently depressed category, and as Table 12 shows, this "persistency percentage" was inversely related to the size of the labor market, being highest in the smallest areas and lowest in the very large markets. Indeed, over half of all the smallest

26. Excess unemployed means the number of employees who were unemployed at a given point in time and who would be required to be at work before the level of local unemployment fell below 6 percent in all qualified areas.

Table 11. Distribution of Qualified Areas by Unemployment Rate, February 1967

Rate of unemployment (%)	Number of areas	Labor market type		
		Major	Smaller	Smallest
Less than 6	2	2	0	0
6 – 7.9	304	13	43	248
8 – 9.9	128	2	18	108
10 – 11.9	65	0	6	59
12 – 14.9	39	0	4	35
15 or more	34	0	0	34
Total	572	17	71	484

areas remained in the persistent and substantial unemployment category. Of course, this inverse relationship could be explained in terms of the higher proportion of the very smallest areas with very heavy unemployment in 1963, which would have had to be pro-

Table 12. Number and Percentage of Areas with Substantial and Persistent Unemployment between 1963 and 1967

Type of area	Number of areas		Persistency percentage
	1963	1967	
Major	15	5	33.3
Smaller	92	34	35.9
Smallest	421	224	56.8
Total	528	263	50.0

foundly affected by national expansion before they were removed from the distressed area group. The evidence does not support this supposition. (Table 13.) As one might expect, the persistency percentage was highest for all areas, regardless of labor market size, which had heavy unemployment in 1963, but for all initial levels of unemployment up to 10.0 percent the smallest areas had a significantly higher "persistency percentage" than other labor market types.

All of this can be summarized in a few words. Currently the scale of distress as measured by the rate of unemployment is inversely related to the scale of the labor market. Thus the really distressed local economies tend to be in the smaller and especially the smallest labor market categories. Moreover, this inverse relationship has tended to become stronger in recent years, precisely because the smallest areas have responded least well to the overall growth in the economy.

Table 13. Persistency Percentage by Labor Market Size and Initial Rate of Unemployment, 1963

Rate of unemployment (%)	Persistency percentage		
	Major areas	Smaller areas	Smallest areas
Less than 6.0	0	7.1	31.5
6.0 – 7.9	14.3	15.2	34.5
8.0 – 9.9	50.0	47.6	62.9
10.0 – 11.9	50.0	61.5	63.5
12.0 – 15.9	0	85.7	75.6
15 or more	0	100.0	80.5

The Socio-Economic Characteristics of Low Income Areas

This is the most typical form of distressed area, which bears all the signs of structural underdevelopment, resource immobility, and personal deprivation. Basically, there are two varieties in this low income grouping. The first type is the rural area, which has lost population for many years as output has declined and/or capital has been substituted for labor. The remaining working population, which contains a high proportion of Negroes, is badly educated, badly housed, and in some instances not only deprived of adequate material sustenance but deprived of civil rights as well.

The second type of area has all the above characteristics but in addition suffers from severe and persistent unemployment. These are the areas of acute depression, where the local economic structure, whether it is based on coal mining, forestry, or agriculture, has collapsed. In addition, these areas often are mountainous, topographically unsuitable for further development and, more often than not, locationally isolated. In some cases human deprivation is accompanied by environmental spoliation and pollution.

Area Distress: Its Current Scale

Area distress cannot be measured solely in terms of economic statistics. Many individual writers and official reports bear eloquent testimony to the fact that low incomes and persistently high unemployment are simply the outward signs of deprivation which is not only material but also personal and psychological.

Even economic measures of the dimensions and costs of distress are, of necessity, very crude. There is no satisfactory way of indicating the economic loss to the nation from the existence of 150,000 families within qualified area who have totally unsatisfactory incomes.[27] With heavy unemployment areas, the theoretical costs to the nation can be measured by the output foregone and the transfer payments which are necessary to support the unemployed. However, even such a simple measure as this is fraught with difficulties. Area redevelopment policies primarily aim at eliminating structural unemployment. However, in any given local unemployment pool there are those who are unemployed for other than structural reasons. Some are frictionally unemployed and en route to suitable jobs within an accessible labor market; some are unemployable in the sense that their skills, location, work aptitudes, physical or mental abilities make them unable to adjust to normal labor demand, and finally there are those affected by inadequate aggregate demand. The residue represents the structurally unemployed, a category containing workers with skills which must be adjusted to changing demands but for whom jobs exist in distant labor markets and those who are unemployed because complementary skills are in short supply.

For any given volume of local unemployment above the frictional level, this structural component will be changed according to changes in final demand, in production functions, and in investment decisions which result in local plant closures. Thus in any given labor market an accurate measure of structural unemployment would be excessively difficult.

A second unresolved issue is whether the current statistics relating to area unemployment are sufficiently comprehensive. There is increasing evidence that spatial concentrations of high and persistent unemployment can exist within a metropolitan labor market which, as a whole, is operating at a full employment level. A recent survey of 14 slum areas in 13 major cities showed that with a national unemployment rate of 3.7 percent the average for the slum areas was 10 percent. In addition, a closer examination of activity rates and of the precise nature of employment undertaken by those at work resulted in the formulation of a "subemployment index" which indicated that one out of every three residents had a serious unemployment or underemployment problem.[28] It is not unreasonable to

27. R. M. Rauner, *Regional and Area Planning: The E.D.A. Experience* (EDA, Mar. 1967), p. 17 a.

28. U.S. Department of Labor, Manpower Administration, *Area Trends in Employment and Unemployment*, May 1968, p. 8.

assume that a large part of this unemployment and underemployment is structural. Thus the rapid growth of job opportunities in the suburban areas of SMSA's has not resulted in a radical reduction in the central city unemployed largely because of job discrimination, suburban housing discrimination and inadequate public transportation between the core and the suburbs.

The third difficulty is that a persistent deficiency of jobs may result in involuntary withdrawals from the labor force, so that there are hidden reserves of labor in the qualified areas which do not appear in unemployment figures. The limited evidence on this[29] suggests that those males in the youngest age group (14–24) and in the oldest group (60 years and over) may have a strong tendency to temporarily leave the labor force if local employment conditions are unfavorable.

Even without precise measurement of these hidden facets of the problem, it is clear that the distressed area malaise is a significant national problem. In fiscal 1966 the average unemployment in qualified areas was over one and a half million, which represented 46 percent of the national volume of average unemployed for that year. If this unemployment rate had been reduced to 6 percent in every distressed area without any net reduction in labor force size, then 132,000 employees (0.2%) would have been added to the employed labor force of the country, with a consequent reduction in national unemployment of about 4 percent. This addition of 132,000 employees would have meant an increase in the gross national product of $1.3 billion and a reduction in unemployment benefits of at least $68 million.[30]

Area Distress: Its Future Scale and Incidence

Any predictions about the scale and geographical incidence of area distress inevitably is fraught with uncertainties. For example, an analysis of major and smaller areas of substantial unemployment in the ten year period between 1957 and 1966 shows significant shifts in geographical incidence (Table 14). Approximately the same number of areas were designated by the Department of Labor[31] in

29. M. Segal, *Unemployment in Distressed Areas* (U.S. Department of Commerce, Area Redevelopment Administration, 1964).

30. These are very crude estimates based on the assumption that one million employed workers produce $10 billion worth of goods and services. The unemployment benefit calculation is based on the 1966 figure of $1.78 billion paid out as benefit.

31. *Area Trends in Employment and Unemployment,* several issues.

both years, but as the table shows, two of the eight regions, the Great Lakes and the Southeast, experienced a large drop in incidence whereas in two other regions, the Southwest and Far West, the reverse occurred.[32]

Table 14. Major and Smaller Areas of Substantial Unemployment, 1957–66

Region	Number of areas	
	1957	1966
Northeast	7	7
Middle Atlantic	15	11
Great Lakes	11	2
Southeast	32	19
Plains	2	2
Southwest	2	7
Mountain	0	1
Far West	0	16
Total	69	65

However, although it is extremely difficult to predict exactly where distress will occur, it is possible to describe the general characteristics of those areas which are likely to suffer from concentrated distress. A detailed projection to 1975 of employment, population, and income by region and county has been made by McGuire and Harris with the following crucial conclusions.[33] They stress that if interregional migration patterns between 1965 and 1975 are on the same scale and have the same direction as in the period 1950 to 1960, then we can expect an equalization of *regional* unemployment rates. Thus, in general, the operation of the market will generate a loss of population in those areas where the demand for labor is less than the available supply and an inflow of population where the reverse situation applies. However, they also conclude that employment growth will be distributed unevenly across national space, with some counties showing much faster than average growth and

32. Some of these regional shifts were heavily weighted by changes within particular states, with sharp declines in Pennsylvania, Indiana, Michigan, Kentucky, and North Carolina and the reverse in New York State and in California.

33. M. C. McGuire, *Program Analysis and Regional Economic Objectives,* Papers on the Analysis and Evaluation of Public Expenditure, Joint Economic Committee, 91st Cong., 1st sess.; and Curtis C. Harris and M. C. McGuire, "Planning Techniques for Regional Development Policy," *Journal of Human Resources,* Vol. IV (1969), pp. 466 ff.

some 7 percent (394) actually experiencing an absolute decline in the demand for labor. Perhaps the most important finding is that the projected employment growth rate varies according to the size of the county, with the counties under 50,000 and over 500,000 likely to grow significantly less fast than the counties in the size range between these figures.

APPENDIX: STATE COMPONENTS OF REGIONS

Northeast
- Maine
- New Hampshire
- Vermont
- Massachusetts
- Rhode Island
- Connecticut

Middle Atlantic
- New York
- New Jersey
- Pennsylvania
- Delaware
- Maryland and D.C.

Great Lakes
- Ohio
- Indiana
- Illinois
- Michigan
- Wisconsin

Southeast
- Virginia
- West Virginia
- North Carolina
- South Carolina
- Georgia
- Florida
- Kentucky
- Tennessee
- Alabama
- Mississippi
- Louisiana
- Arkansas

Plains
- Minnesota
- Iowa
- Missouri
- North Dakota
- South Dakota
- Nebraska
- Kansas

Southwest
- Oklahoma
- Texas
- New Mexico
- Arizona

Mountain
- Montana
- Idaho
- Wyoming
- Colorado
- Utah

Far West
- Washington
- Oregon
- California
- Nevada

5

THE EVOLVING FEDERAL STRATEGY— PROGRAM IMPLEMENTATION

Judged by administrative criteria, the first two years of EDA activity were a clear success.[1] The administration's actual requests for agency funds, far short of the original legislative authorizations, were met almost in full by congressional appropriations.[2] By the end of June 1966, many of the 400 personnel in the Area Redevelopment Administration had been absorbed into EDA, and over 500 new personnel recruited. Roughly 40 percent of this total staff was employed in seven field offices spread throughout the country.[3] At the end of fiscal 1967, five separate Regional Action Planning Commissions covering parts of nineteen states had been established, and twenty-four multicounty development districts covering 190 counties in twelve states had been officially designated.

The program appeared to be popular with designated areas. No fewer than 3,600 applications for assistance were received, with project requests totaling over $2 billion.[4] Certainly by no means all of these applications were "serious starters." A large number were hold-overs from the ARA and Accelerated Public Works programs, and many of those were ineligible on the basis of the new legislative criteria, some had been directed to the wrong agency, and others were out of date. Nevertheless, in the first six months of the EDA program over 1,000 new applications were filed.[5] By the end of fiscal 1967, over 2,500 of the total applications had been processed,

1. This chapter will focus largely on the first two years of EDA activity, that is, 1965–67, though critical changes in direction in 1968 will also be analyzed.

2. The combined requests in fiscal 1966 and 1967 for EDA funds, including the regional programs but excluding Appalachian authorizations covered by the 2nd Supplemental Act of 1965, were $675 million. Of this $628 million was appropriated and a rather smaller amount obligated.

3. The seven area offices are in Portland, Maine, Wilkes-Barre, Pa., Huntington, W.Va., Huntsville, Ala., Duluth, Minn., Seattle, Wash., and Austin, Tex.

4. U.S. Department of Commerce, EDA, *Regional Economic Development in the United States*, Oct. 1967, Appendices, App. B, Table I.

5. *Report of the Hearing before Special Subcommittee on Economic Development Programs*, House Committee on Public Works, 90th Cong., 1st sess., Apr. 11, 1967, p. 74.

1,300 of them accepted and 1,200 denied.[6] A small number of projects were withdrawn voluntarily, so that the project-pending backlog had been reduced to 700 from the previous year's total of 1,700. As far as can be ascertained from appropriation hearings and from special congressional subcommittee investigations, these approvals and denials were accomplished with a minimum of allegations of mismanagement.[7] In total, Federal finance promised for approved projects amounted to $563 million, $150 million actually having been disbursed through fiscal 1967. Including the non-Federal share of project financing, total commitments reached almost $1 billion.

Most of the Federal funds were distributed for public works projects, and particularly for water supply and sewage installations, but as Table 15 shows the amounts committed to other programs amounted to over $100 million. Every state received some assistance, and a high proportion of the designated areas achieved at least one project approval. For example, in fiscal 1966, the number of qualified areas was usually somewhere between 1,200 and 1,300 and typically at least 80 percent of these areas were designated. Of this latter figure, representing roughly 1,000 areas, almost one-third re-

Table 15. EDA Investment by Program Headings, 1966 and 1967

Program	Cumulative EDA investment
Public works	$431,579,000
Business loans	93,925,000
Working capital guarantees	10,115,000
Planning grants	4,786,000
Technical assistance	17,950,000
Research	5,392,000
Total	$563,747,000

Note: From U.S. Department of Commerce, Economic Development Administration, "Directory of Approved Projects," June 30, 1967. p. iii.

ceived some form of project assistance. In fiscal 1967, project coverage was even more spatially extensive. With many holdover areas

6. *Regional Economic Development*, Oct. 1967, Appendices, App. B, Table I.

7. In fact most of the criticisms arose from cases in which the agency had to interpret a highly complex section of P.L. 89–136, which sought to prevent financial assistance from being given to companies which would increase production of goods and services in conditions where there was insufficient demand for such expansions, or where existing efficient capacity was underutilized.

Table 16. Distribution of Approved Projects by Local Population Size, Fiscal Year 1966

Population size of area	Percentage of all EDA qualified areas		Public works		Business loans		Technical assistance	
	Number	Percent	% projects	% EDA funds	% projects	% EDA funds	% projects	% EDA funds
1 – 1,499	11	0.9	21.9	8.8	35.5	45.9	16.2	10.2
1,500 – 2,499	30	2.4	10.5	5.6	9.7	9.5	5.4	1.0
2,500 – 4,999	86	6.8	15.8	10.2	8.1	7.3	10.8	3.0
5,000 – 9,999	242	19.1	18.5	19.1	19.4	19.9	14.9	8.9
10,000 – 14,999	207	16.3	8.6	4.3	4.8	3.0	2.7	2.2
15,000 – 24,999	285	22.5	6.4	4.7	4.8	3.2	6.7	13.0
25,000 – 99,999	330	26.0	10.0	19.3	11.3	8.9	18.9	21.4
100,000 – 499,999	63	5.0	5.5	19.7	6.4	2.3	12.2	14.2
500,000 and above	13	1.0	2.8	8.3	0	0	12.2	26.1
Total	1,267	100.0	100.0	100.0	100.0	100.0	100.0	100.0

Source: Data taken from qualifications as of Dec. 1965, EDA, "Maximum Grant Rates," Dec. 1965, and *Regional Economic Development*, Oct. 1967, appendices, App. B, Table III. The technical assistance data cover 46% of all projects.

disqualified, an average of 1,000 areas were qualified, and once again approximately 80 percent of these had an approved overall economic development program which guaranteed designation. No fewer than 45 percent of these areas (382) received project assistance, with just under 300 gaining a public works project.

There are no data on the distribution of 1967 projects by population size of designated areas, but the 1966 figures suggest that both the public works and business loans programs were heavily focused on the smallest distressed communities, whereas technical assistance was more evenly distributed throughout the range of population size. Indeed, as Table 16 indicates, approximately half of the projects for public works and business loans went to areas of less than 5,000 population, whereas the relevant proportion for technical assistance was only one out of three. This distribution of projects was markedly different from the frequency distribution by size of all qualified areas. Thus the very small areas with less than 3,000 population gained a disproportionately large percentage by number and financial share of the public works and business loans projects.

It is unrealistic, at this stage, to indicate the nature and impact of project assistance upon designated distressed communities. In part this is because much of the agency's work is directed towards the stimulation and promotion of active experiments by distressed communities themselves in local economic planning and programing, and this objective has an unmeasurable short-term pay-off.[8] Moreover, even the "hardware" part of the program is new, and disbursements to date are very limited. In any event, for every program of public works there are obvious conceptual and statistical difficulties in isolating the immediate local cost and benefit streams generated by specific public investments from those effects caused by changes in aggregate demand and by competition. It is even more complicated to quantify the long-term income and employment multiplier effect of public investment and above all to identify the long-term relationships between net investment and the attraction of net private capital which otherwise would not have been invested locally. There is no question, however, that over the two years of agency operation, the combined effect of an aggregate demand increase, area-specific EDA investments, and the disqualification of many "holdover" areas caused a marked reduction in the statutory dimen-

8. It is noticeable, however, that EDA did not establish formal administrative mechanisms for evaluating these experiments in the first three years of agency activity, though ad hoc evaluations were continuous.

sions of the distressed area problem. (Table 17.) Thus the number of qualified areas fell by over 20 percent to under one-third of all areas, and this in turn reduced the proportion of the total labor force and total unemployed in qualified areas to 16 percent and 31 percent respectively.

Table 17. Estimate of EDA Program Dimensions

Some key indicators	U.S.		EDA qualified areas	
	FY 1966	FY 1967	FY 1966	FY 1967
Number of areas	3,100	3,100	1,280	1,000
Civilian labor force (× 1,000)	75,635	77,000	19,400	12,070
Number unemployed (× 1,000)	3,456	3,000	1,591	930
Percentage rate unemployment	4.6	3.9	8.2	7.7

Note: From *1968 Appropriation Hearings,* House of Representatives, Part 3: Department of Commerce (U.S. Government Printing Office, 1967), p. 181.

The New Priorities

Despite the growing momentum and apparent success of the program, by early in the second year of operation the chief executives of the agency had initiated a fundamental reappraisal of policy. In part this was a natural response to the end of a phase in which priority had rightly been given to the creation of a complex administrative system, the appointment and training of personnel, and above all the initiation and processing of thousands of project applications. It also reflected an explicit awareness that with the significant reduction in the number of distressed areas due to the disqualification of many holdover areas, the problem to be solved was now at a more normal level. However, the real seeds of impending change lay far deeper than either of these factors and can be traced to a growing awareness within the agency that funds must be used more effectively if the condition of the distressed areas was to be ameliorated to any degree. This conclusion was based upon three observations. Given the number of designated areas, the size of the backlog of pending projects, and the absence of any anticipated increase in agency staff, it was entirely realistic to expect that the annual number of project submissions would exceed, by a wide margin, the annual processing capability of the agency. Secondly, if the agency continued to react to project requests as and when they arose and did not prescribe in advance the number of projects which could be handled, then there would always be a tendency to underload or overload the agency's resources at particular points in time. The inefficiency of such a system would be most apparent in the months immediately prior to the end of the fiscal year, and especially if a large proportion of appropriated funds had not been obligated so that the agency was forced into making hasty decisions in order to show that all the appropriated funds were truly necessary. The third problem was the most complex of all. If annual appropriations continued to be of the order of $300 million, then the agency could expect to fund about 800 new projects each year[9] and would have to reject several hundred claims. There was therefore a clear need to establish a set of criteria on which to base the allocation of scarce resources to competing demands.

Of these problems, the questions of how to moderate the overall number of requests for assistance and how to phase their flow were both amenable to relatively simple administrative control. Accordingly, during the course of fiscal 1967 several steps were taken both to diminish and even out the flow of requests and to enlarge processing capability. Field officers were instructed to scrutinize projects carefully before advising on submission and to cooperate with other Federal agencies in directing applications to the correct Federal agency. The processing of business loans, which involved close liaison with the Small Business Administration, was simplified, and a determined attack was made on the backlog of unscrutinized projects. In addition the agency made no efforts to extend its responsibilities in two areas which would have greatly increased the number of potential claimants upon scarce funds. Despite considerable congressional pressure, the agency consistently declined to seek amendments to the legislation which would have permitted assistance to distressed areas of less than the size prescribed by the 1965 act, and it insisted on using the loss-of-population criteria in conjunction with low income criteria. For a considerable period, the possibility of using EDA public works and business loan funds in city ghettoes was under active consideration, but a decision was finally taken not to seek amendments to the legislation, and to concentrate assistance on the existing problem set.

9. In the first two years of agency activity, appropriations averaged $308 million, with $290 million for public works and business loans and $18 million for technical and planning assistance. The crude average cost for the first type of project was $600,000, so that 480 projects could be funded, and for the second type, $50,000, so that 360 could be funded.

There were, however, no simple solutions to the problem of how to select investment priorities. As we have already noted, the legislation of 1965 provided no unequivocal guidance on this, and the experience of the agency in the first year of operation simply confirmed that the legislators had selected very general program objectives and no firm basis on which to base investment priorities. The net result was a year of constant reevaluation within the agency of how the objectives of the act were to be realized. Two questions were at the center of this debate, one relating to the precise nature of the objective of the agency in aiding distressed areas, the second relating to the benefits of concentrating investment in selected areas.

It was generally agreed within the agency that the key indicator of success in assisting designated areas was the number of permanent jobs which arose as a direct or indirect result of Federal funding. However, the critical question was the degree to which the impact per public dollar invested should be represented in the evaluation procedure. Three alternative solutions were propounded. The first was to use the benefits created per public dollar as the sole criterion and to rank projects in descending order of benefit contribution. Another method was to use the efficiency criterion described above in ranking projects but to supplement it with a "handicap." This could be based on the degree of distress as measured by unemployment and/or median incomes in the area in question, and also on subjective evaluations by field officers and reviewing officials as to the worth of the project in terms of its experimental value, demonstration qualities, importance to planning activities in the community, and so on. The final method was to select the "worst" areas, as indicated by high levels of unemployment, high net migration, and low median incomes, and to concentrate Federal spending within such areas.

The choice of the last of these methods, that is the "worst first" strategy, was occasioned by a comprehensive analysis of the first year of agency activity and the clarification of the basic weaknesses in a system based upon grass-roots project generation and Federal government reaction. A major criticism of the Area Redevelopment Program has been that communities received insufficient assistance from the Federal government despite the appropriations for local economic development and vigorous local efforts in creating overall economic development programs. The executive staff of the EDA therefore decided that the new program must really provide every opportunity for as many communities as possible to obtain Federal assistance, and secondly that it must be characterized by efficient handling of project requests. Accordingly, the field officers of EDA were encouraged to initiate interest in the new program and to exhort local development committees to submit project requests. As we have already noted, this led to a huge number of project submissions and ultimately to efforts to restrain and even out the flow. But the really crucial point is that there was a marked tendency for the areas with the least severe conditions of distress not only to submit projects but to have these projects approved. It is not difficult to see why this occurred. Given the overall policy of the agency, a realistic objective for any individual field officer was the maximization of the number of successful project requests. Hence, one obvious strategy was to make contact with *existing* community organizations which were generally recognized as representative and had high-caliber officials and members who could quickly generate viable project requests. Such bodies are most likely to be operative in communities which have reasonable prospects for redevelopment.

The method used by the agency to evaluate projects may also have favored areas which had the ability to articulate precise project requests, for although the evaluation method included some subjective judgments on the importance of any given project to the area, the principal considerations were whether the project conformed to the overall economic development program and whether it was likely to make a satisfactory direct and indirect contribution to the creation of permanent local jobs.

The result of this policy was not only that a very large number of areas received some project assistance but also that many relatively prosperous areas had projects funded. Indeed it was discovered that 110 of the 340 separate areas which received some help were actually disqualified at the 1966 annual review on the grounds of having an unemployment level of under 6 percent. Now since EDA project assistance was largely in the form of public works grants, which are likely to have an impact upon local activity levels in the long term, it was concluded by the chief officers of the agency that the major cause of the reduction in unemployment was the increase in the aggregate level of effective demand rather than direct Federal action. A detailed statistical analysis which was made within the agency tended to confirm this conclusion since it showed that areas which initially had the lowest rates of unemployment were much more likely to be disqualified than areas which initially had high rates of unemployment.[10] (Table 18.)

A second result of this overall policy was that very few areas received anything approaching a "critical mass" of investment, the typical allocation being one or two projects. Once again, the reasons for this outcome are not difficult to discern. For administrative convenience, the agency put its funding functions under a number of separate committees, so that requests for public works or technical assistance or business loans were processed by different parts of the agency. Inevitably this meant that the chances of composite funding were considerably reduced.

Table 18. Unemployment-Qualified EDA Areas Terminated in June 1966

1964 unemployment rate	Qualified November 1965– June 1966 [a]	Terminated June 1966 [b]	Percent terminated
12.0 or higher	113		
10.0 – 11.9	111	2	1.8
8.0 – 9.9	247	16	6.5
6.0 – 7.9	277	103	37.2
Total, all areas	748	121	16.2

Source: Bureau of Employment Security, U.S. Department of Labor, Sept. 1966.

Note: National average unemployment rates: 1964, 5.2%; 1965, 4.6%.

[a] Areas qualified in 1965 with 1964 average rate of unemployment of six percent or higher.
[b] Areas with 1965 average rate of unemployment under six percent.

The method chosen to overcome this scattered dispersal of agency resources was courageous and ingenious. The fundamental premise was that if the agency continued to fund projects in areas which would soon fall below the qualifying threshold because of aggregate demand increases, then there would be a real cost to the Federal government. This cost would be the opportunity foregone of using these funds in areas where changes in aggregate demand would not affect local employment levels. Thus with all distressed areas arranged in a rank-order based on the extent of the gap from the distress level threshold, the opportunity cost to the Federal govern-

10. Source: *Regional Economic Development,* Oct. 1967, Part II, p. VI–31. Note that Table 18 shows 121 areas were terminated whereas the figure above refers to 110 out of 340 areas. The figures in the table relate to all areas terminated whether they received EDA assistance or not.

ment of reducing this gap would decrease exponentially. In consequence "the largest relative cost of gap reduction is to be found at the bottom of the list. The smallest relative cost of gap reduction is to be found at the top in the 'worst' area. Hence EDA's most expensive program would result from any deliberate policy to maximize the amount of employment and income per dollar of EDA investment without regard to the ability of an area to respond to national growth without assistance."[11]

Given this view, the general policy objective was refined into seven distinct goals, that is, one for each of the seven programs of qualification for project assistance. Therefore, within the constraints of appropriated funds, and a distressed area set comprising the designated areas, the agency committed itself to

"(1) minimize the maximum area unemployment rate for 'substantial' unemployment level areas;
"(2) minimize the maximum unemployment rate for 'substantial and persistent' unemployment level areas;
"(3) maximize the minimum area median family income for low income areas;
"(4) maximize the minimum median family income for low income/population loss areas;
"(5) provide at least one new job opportunity for every two Indian area families;
"(6) provide a sufficient number of new employment opportunities to offset anticipated job losses in eligible areas threatened by a 'sudden rise' in unemployment due to a major employment shutdown;
"(7) reduce the most recent annual average unemployment levels by ten percent in 'Mink' areas."[12]

There can be no doubt that the choice of the worst-first strategy provided the basis for improving the operating efficiency of the agency in several distinct ways. The single most important benefit was the selection and explicit acknowledgment of one goal from among the welter of potential goals suggested by the legislation of 1965. This not only focused the energies of EDA staff at head-

11. *Regional Economic Development*, Oct. 1967, Part II, p. VI–30.
12. *Regional Economic Development*, Oct. 1967, p. VI–23. "Mink" areas are those which qualify as distressed on the grounds that no other area within a given state is eligible under the six other methods of qualification. They were called "Mink" areas after the representative from Hawaii, Congressman Mink, who originally suggested this method of qualification.

quarters and in the field, but assured local planning agencies in selected areas of continued Federal involvement with their development problems. Both of these factors permitted the agency to move from a passive role vis-à-vis a multitude of distressed areas towards an active scrutiny of which EDA subsidies were relevant aids for the revitalization of specific communities. Indeed, the agency fully recognized that this type of discrimination provided the opportunity for public investment not only in "critical masses" but as part of a sequence of public investment commitment over time. This was a critical change in direction and one which presaged a subtle shift in the shared responsibilities of the parties actively seeking the revitalization of distressed areas. Up until this time it had always been assumed that the distressed communities themselves would set their own development goals, chart their own strategies and implement their own plans. The Federal role in all this was limited to the provision of advice on redevelopment planning and the advancing of "seed capital." This new policy meant that the agency assumed responsibility for setting the development target (i.e., the specified gap reduction) and committed its resources towards this end, albeit with the assistance and cooperation of local citizens who would remain as the prime implementers of the agreed strategies.

A second implication of the new policy was the potential dilution of the political content of fund allocation decisions. With the stress laid upon general allocations to seven distinct programs, and within these programs, upon specific allocations to particular areas on the basis of the resources required to attain a targeted reduction in gap units, the scope for political activity to determine the spatial content of fund allocation was severely constrained. Indeed the use of redevelopment area targets permits the agency to *predict* aggregate fund allocations for any spatial grouping thought desirable on political grounds. For example, the funds to be allocated in any fiscal year can be aggregated by region, by state, by multicounty grouping, and so on. (Table 19.)

All of this relates to another advantage of specifying a unique goal and programing the use of funds, namely, the possibilities over the long run of measuring the effectiveness of the agency's use of resources as expressed in a cost per gap unit reduced across the range of assisted areas. Indeed it is only within such a framework that budgetary performance measures have any meaning. Up until this stage, and despite several attempts to standardize the objectives for

Table 19. Programs and Gap Units

Qualification ranking	*Program 1* Substantial unemployment		*Program 2* Substantial & persistent unemployment		*Program 3* Population loss/low income		*Program 4* Low family income		*Program 5* Indian areas	*Program 6* "Sudden rise"		*Program 7* "Mink" areas	
	Unemp. rate (an. av. %)	Gap (an. av. no. pers.)	Unemp. rate (an. av. %)	Gap (an. av. no. pers.)	% fam. below 50% nat. median	Gap (no. fam.)	% fam. below 40% nat. median	Gap (no. fam.)	Gap (no. addtl. jobs reqd. at rate of 1 for every 2 fam.)	Unemp. rate (an. av. %)	Gap (an. av. no. pers.)	Unemp. rate (an. av. %)	Gap (no. pers. to red. unemp. by 10%)
Highest	White Sulphur Springs, Mont.		Wade Hampton, Alaska		Wolf, Ky.		American Samoa		Navajo Res., Ariz.	Brooklyn Navy Yard, N.Y.		Bristol, Conn.	
	8.8	29	31.4	162	78.6	408	86.1	930	8,430	13.0	2,500	6.4	130
2d highest	Merced, Calif.		Bethel, Alaska		Owsley, Ky.		Puerto Rico		S.E. Res., Alaska	"		Albert, Vt.	
	8.7	1,006	29.4	264	78.3	351	70.5	46,775	1,790	"	"	5.8	10
Middle	Crescent City, Calif.		Brownsville, Ky.		Greene, Ala.		San Jacinto, Tex.		Cook Inlet Res., Alaska	"		Sussex, Del.	
	7.4	92	18.1	321	72.2	624	61.1	172	1,240	"	"	4.6	130
Next-to-lowest	The Dalles, Oreg.		Richlands, Va.		Karnes, Tex.		Northampton, N.C.		Duck Valley Res., Nev.	"		Monroe, Iowa	
	6.0	10	6.0	12	50.2	10	50.2	10	100	"	"	4.4	16
Lowest	Lowville, N.Y.		Hibbing-Virginia, Minn.		Walnut Ridge, Ark.		Gates, N.C.		Ramah Res., N. Mex.	Marshfield, Mo.		Guam	
	6.0	10	6.0	10	50.1	431	50.0	10	100	7.3	59	2.2	39
Total	6.7 (av.)	23,724	9.7 (av.)	105,305	60.3 (av.)	99,342	55.2 (av.)	51,329	26,501	10.1 (av.)	2,559	4.9 (av.)	2,578

Source: R. M. Rauner, *Regional and Area Planning: The EDA Experience* (EDA, Mar. 1967), pp. 17a, 17b, 18, 19.

Note: Unemployment rates relate to 1966. In programs 1, 2, and 6 the gaps shown against each area indicate the number of unemployed workers who would have to be employed before that area was not qualified to receive EDA assistance. Similarly in programs 3 and 4 the gaps relate to the number of families that would have to have their incomes raised, so that no family would have less than 50% (program 3) or 40% (program 4) of the national median income. In programs 5 and 7 EDA chose as their targets one net job created and a 10% reduction in the local rate of unemployment, respectively. Under each program, the totals show the number of workers or families requiring assistance and, in some broad sense, the relative magnitude of the different types of distressed area problem.

Federal project assistance, it was not apparent that budgetary performance could be measured area by area or by type of subsidy.

It is indisputable that the selection of the worst-first strategy brought a much needed focus to the agency's activities. Nevertheless there are legitimate grounds for doubting whether this was the correct focus. The justification for concentrating upon the worst areas was not that the benefits of new jobs and incomes per dollar of public subsidy were higher in such areas. Thus it was assumed that the marginal utility of a given job, at a given income, was invariant with the level of distress. In fact the defense of the strategy was based on the assumption that one-third of agency funds had been wasted in the first year of operation because there had been no explicit system which prevented investment in areas with a high probability of disqualification due to a general growth in the economy; and, secondly, it was based upon a belief in the efficacy of EDA subsidies in aiding the economic revitalization of the most distressed areas.

Now the first defense is, in many ways, rather curious. Although it is undeniable that almost one-third of all assisted areas were ultimately disqualified and it is likely that very few of these areas moved below the qualification threshold because of EDA activity, it does not follow that EDA funds were wasted. On the contrary it is reasonable to assume that some of those areas which were assisted have a better chance of remaining disqualified precisely because this assistance has helped remove deficiencies in the local economy which might have caused future disequilibrium. Even if we cannot assume that this argument has validity, the supposition that funds were wasted is at least partly dependent upon the typicality of the years covered by the analysis.

Much depends upon whether the future growth of the economy can be expected to cause as large a number of areas to fall below the qualifying threshold as in 1964–65. In fact this latter period was one of exceptionally rapid expansion, with the administration using fiscal and monetary measures to raise the level of effective demand in order to combat a persistently high rate of unemployment. More recent years have seen a less sharp rise in activity levels and a considerable reduction in the annual number of areas being disqualified because of exogenous growth. Thus, even if the agency continued to assist areas on broadly the same principles as before, a declining proportion of all areas would be disqualified because of extra-agency economic influences, and therefore a declining number of subsidies would be "misallocated."

However, a more important question is whether, as the worst-first strategy implies, the public cost of gap reduction actually does increase exponentially as the qualifying threshold is approached. We must be clear about the exact meaning of this claim. This use of cost is not the economist's "opportunity cost." Thus "cost" in this context is actually a statement of probability relating to the likelihood of any particular area's being affected by national growth effects. To concentrate on the worst areas, it is argued, generates the least risk of subsidies being used in areas which are going to be disqualified anyway.

The first point to note about this argument is that there was no marked difference in the termination of areas with initial rates of unemployment between 8.0 and 9.9 percent as compared to those with rates of 10.0 to 11.9 percent.[13] Hence there is no overwhelming reason why assistance should be concentrated in the very worst areas instead of in any area which is operating at 8 percent unemployment or above. The second point is even more critical. While it is entirely legitimate to choose the areas with the most severe conditions of distress as the prime target for agency activities, we have continually stressed that many of these areas may be entirely unsuitable for structural transformation, even though the agency actively pursues their development. Certainly the agency itself qualified its commitment to aid the designated worst areas in four distinct ways:

1. "In applying its strategy, EDA recognizes that not all areas are equally capable of absorbing EDA investments effectively. Therefore, it is necessary to determine the scope and amount of EDA activity in terms of the areas' potential for efficient use of resources."

2. "Projects and investments are most favored that most directly affect the unemployed or underemployed. EDA does not accept the 'trickle down' theory of economic improvement, that it is enough to promote economic growth and let the benefits trickle down to the most distressed people."

3. "Major consideration is given to continuity of local efforts to maximize the effective use of all available resources. This requires, among other things, satisfactory planning, community involvement

13. In the first grouping (8.0 to 9.9%) 16 of the 247 areas, that is, 6.5%, were terminated, whereas in the second unemployment grouping (10.0 to 11.9%) 2 out of 111 areas or 1.8% were terminated. See Table 18.

and support for the development effort, and effective implementation capability."

4. ". . . a [project] selection or strategy for a single area cannot be taken without first having examined the area's relationship to its geographic and economic settings. For . . . it may well emerge that a preferred . . . strategy is to concentrate investments on assistance in an adjacent area. This means . . . that each individual area must be arrayed for its possible relationships to a next-higher economic entity, be it another area, a District Growth Center, a District or a Region."[14]

These qualifications might suggest that the agency has not made an unequivocal commitment to investing resources in the worst areas but has simply established a priority system for the consideration of redevelopment area problems. And yet, the real test of this is whether EDA will be able to examine the problems of the worst areas and then to state explicitly and publicly that it does not possess the resources to assist some of these areas. For a variety of reasons discrimination of this kind is unlikely. In part this is because the organizational directives become diffused and imprecise if they are expressed in conditional language. Notably, when first explaining the new policy to legislators, the chief executive of the agency stressed that EDA would "stick with the worst areas. . . ."[15] In any event, the agency may find that it is unable to desist from aiding communities which have shown a genuine interest in and involvement with the development of their locality. There is a marked difference between rejecting a project from an area and explicitly recognizing that certain areas, though receiving priority consideration, are nevertheless unsuitable for assistance. The evidence for 1967 suggests that this possible concentration upon areas which may have a very low potential for economic development did in fact begin to take place. When comparing the results in the last three months of fiscal 1966 with the three months in fiscal 1967, the agency concluded:

> "Thus implementation of the strategy brought more than a doubling in the number of projects and nearly double the amount of money in the worst areas. . . . The opposite picture appears for those eligible areas where the problems are least severe . . . [with] more than a 50 percent reduction in projects and a 70 percent

14. Rauner, *Regional and Area Planning*, p. 21.
15. *Regional Economic Development*, June 1967, p. 1.

redirection of funds. . . . In Fiscal 1968 the Agency will continue to emphasize the 'worst-first' policy."[16]

In the event, the agency appeared to have second thoughts about the effect that the worst-first approach would have on the development district program, so that by 1968 the two approaches were in a state of uneasy coexistence.

The Development District Program

The objectives of this part of the EDA strategy are clear enough. Thus the goals are to create

(a) area-wide (multicounty) planning institutions, staffed by professional planners, who will crystallize area development goals and seek the implementation of strategies for economic development with the active participation and involvement of county governments, private companies, and interested lay individuals;
(b) explicit public investment programs based upon Federal investment in an urban center of some size and past record of economic growth;
(c) a local institutional framework for coordinating the multiplicity of Federal agencies whose activities have a marked effect on local development;
(d) a permanent planning framework which exists to proclaim the merits of area-wide planning strategies and not solely to gain Federal investment for local schemes.

The achievement of these goals inevitably entails long-term Federal involvement. There is no local governmental structure at the multicounty level, and no past record of planning for such a context,[17] and consequently little likelihood of an intimate understanding by the states of the objectives of such a program. In addition, a long tradition of competitive subsidies by individual counties seeking to attract migrant industry is unlikely to create the conditions for a rapid adoption of a multicounty approach.

The activities of EDA in developing the district program were exemplary. Careful, patient discussions were held with individual states prior to the outlining of boundaries, the selection of centers, and the establishment of planning committees. With a few notable

16. *Regional Economic Development,* Feb. 1968, p. 5.
17. Except in some states where substate economic areas have been defined.

exceptions the states welcomed the program. Wherever possible the district boundaries were made to coincide with state economic areas, and where necessary EDA used some of its planning grants to strengthen state agencies involved in the planning of the districts. Furthermore, in addition to local government representatives, EDA took special care to fill the local planning institutions with representatives of a wide range of interests in the districts as a whole and particularly of the poorer sections of local society. Each of the new planning agencies was given constant support by field officers, through carefully prepared literature which outlined the objectives and methods of local area planning.

There seems little doubt that for a strictly limited cost the activities of professional planners at the multicounty level are likely to encourage a much more careful scrutiny of local public investment strategies.

The key question, however, is whether the program will lead to the kind of growth center strategy envisaged in Chapter 3, which requires explicit discriminatory investment in the development center and an acknowledgment that the economic interests of the hinterland areas must be subordinated, at least in the short run, to the rapid build-up of the center.

For this to occur, several changes are required. EDA must reevaluate the nature of its inducement policy with regard to the centers in two respects. Under current legislation, redevelopment areas are encouraged to join in a district by the offer of a "bonus" grant. Areas which commit themselves to the district actually can receive the highest available ratio of grants to total cost of any given public works project. Such an approach actually encourages the formation of local planning organizations at precisely the time when the whole thrust of Federal activity should be to develop the planning capability and planning jurisdiction of the development district planners. This is not to argue that the economic problems of the individual redevelopment areas should be ignored. We have stressed that a complementary policy of structural analyses and renewal may be required for such areas. But the keynote must be a clearly identified program of investment discrimination in favor of the centers, with individual redevelopment areas being encouraged to join an approach which will generate benefits for the district *taken as a whole.*

A second change is required. The subsidized growth of an urban center is a *process* which impinges on a great number of economic actors, but essentially it requires the subsidized creation of job opportunities in the center and encouragement, by inducements and other means, to move people from the hinterland areas to the center, so as to take these jobs. EDA's present inducements, which probably have a highly indirect effect on industrial costs, are heavily slanted to public works. What is required is a much more direct attack on attracting exogenous capital, perhaps through the provision to companies of settling-in grants, low-interest loans, and subsidized training programs. Equally important are the inducements which could be made available to would-be migrants, such as labor market information, placement advice, removal allowances, and assistance in selling houses in the hinterland area.

The corollary of this approach is a real effort to integrate the many Federal activities which impinge upon the economy of the development center and its hinterland region. EDA's present resources are so slender that it cannot hope to provide the major share of Federal investment. However, it can actively pursue a course of seeking agreement with other relevant agencies on program goals and strategies for each development district. If it fails to achieve such agreements, then the benefits of its own investments may well be canceled or diluted by the activities of other agencies.

A similar point can be made with regard to state planning efforts. Though there is an obvious case for Federal initiation and encouragement of the development district idea in the short run, over the longer term the individual states must accept the dominant role for fostering the development of their districts. This suggests that any efforts by EDA to achieve a coordinated program for the development of each district must proceed only if there is visible evidence of the acceptance by a given state of its prime role.

Summary

In 1965 the Economic Development Administration was faced by a distressed area set which covered more than a third of all the counties in the country, a strictly limited program budget, a House of Representatives which expected a significantly improved political pay-off from a program which had almost been allowed to die, and a legislative commitment which was ambiguous in terms of program goals but explicit on the need for new planning institutions. The early response by the agency was entirely rational on political grounds, the major objective being to encourage local areas to submit project requests which were to be reviewed by EDA in Washington in a rigorous but speedy system of project appraisal. Inevitably, the geographical spread of projects was enormous and very few areas received a critical mass of investment. The signal achievement of EDA in its second year of activity was to recognize the overriding necessity for discovering a priority system of area investment, so that those designated areas with higher potential for self-redevelopment would be given modest assistance, and others with low potential would be rigorously analyzed by Federal administrators for signs of their suitability for concentrated planned public investment. The basic weakness of the agency in 1967 was that it adopted a policy of concentrating its attention upon areas with extremely little chance of successful redevelopment, so that it was likely both to waste project resources and to face political denigration, as development results never seemed to extend beyond planning documents and Federal assertions of sympathetic interest. However, by 1968 it had become clear that this strategy of "worst-first" was likely to compound the difficulties of establishing a spatially focused investment policy (a development district approach) which presents real opportunities for arresting the decline of subregions and of mitigating the social costs inherent in unchecked migration from declining areas to fast growing metropolitan centers. Both strategies coexisted uneasily during 1968.

6

NATIONAL, REGIONAL, AND STATE PLANNING AND THE REGIONAL ACTION PLANNING COMMISSIONS

Title V of the Public Works Act of 1965 together with the Appalachian legislation of that year marked the inception of an entirely new dimension in U.S. regional planning. The underlying philosophy behind both pieces of legislation was a belief that long-term economic planning for lagging regions is relatively ineffective if individual states operate entirely within their own boundaries and Federal-state cooperation is molded by narrowly conceived bilateral deals between individual agencies and individual states. In contrast regional planning which derives its power from high powered research, multistate deliberations, and multiple agency Federal involvement was seen as a genuine answer to the social and economic problems generated by unemployment, underemployment, and persistently low incomes. As a result of this view, six Regional Action Planning Commissions were created, six appointments were made to the entirely new post of Federal cochairman, six separate research and planning teams were assembled, governors in twenty-four states actively participated in the work of the commissions, and President Johnson ordered the establishment of a Federal Advisory Council on Regional Economic Development.[1] How valid is this approach and what can be expected from this massive outburst of institution creation?

Regional economic planning as a continuing process, operating over a designated space, has three distinct, though interrelated component activities. The first is *surveillance* and *diagnosis,* a process which includes the monitoring of current economic performance, the quantification of future demands and supplies of human, natural, and man-made factors of production, and the prediction of how and where weaknesses in the regional economy may be expected given the growth path of the national economy. With the second stage an explicit selection of *social* and *economic goals* is made. Such goals may be precisely quantified or left as general unweighted objectives;

1. Executive Order 11386, Dec. 29, 1967, "Federal Participation in Regional Economic Development."

may be extensive and related to a preferred spatial distribution of population and economic activity or narrowly confined to particular areas or sectors of activity; and finally may be drawn up in terms of a given time period or once again left imprecise. The final component involves the adoption of a range of *public policies* and *activities* thought necessary to achieve the selected goals.

Implicitly or explicitly, the legislation of 1965 suggests four reasons why this process of planning is most effective if it applies to regions which are selected on the basis of space-extensive socioeconomic criteria, rather than political boundaries on their own. The first is that lagging regions do not fit neatly within political boundaries. State 'x' does not necessarily suffer from a different kind or degree of structural malaise from contiguous state 'y,' and the faster economic growth of contiguous state 'z,' which does not fall into the category of a lagging area, may nevertheless be directly affected by the slow growth rate of 'x' and 'y.' Thus problem diagnosis may be possible only if all the relevant political jurisdictions are involved. It also follows that remedial programs and policies must be evaluated within a geographical area which internalizes the bulk of any future cost/benefit streams produced by these programs and policies. Otherwise spill-over benefits and costs may be either overestimated or underestimated, and for many major investments, the individual state is likely to be too small for a valid calculation. Thirdly, apart from increasing the range of possible geographical sites for major investments, the type and range of viable investment projects may be enhanced as a direct result of increasing the number of planners and interested politicians. Thus multistate collaboration may extend the pool of ideas on development possibilities, make possible a comprehensive and permanent data system, and result in the pooling of the financial and executive resources of individual states. Finally Federal involvement may be rationalized insofar as a multistate acceptance of a published regional plan allows individual Federal agencies to gear their regional investments to regionally devised priorities.

However, despite these "legitimate" reasons for establishing regional planning institutions, what is most striking is the enormous gap between the obvious potential of the commissions and their achievements in the years between 1966 and 1968. For example, by the end of 1968 two commissions had actually devised and published overall development strategies and all the others were actively engaged in formulating their plans. And yet there is no evidence that the Federal government changed any of its investment activities to fall in line with these plans. Even more significant is the fact that the President did not seek anything other than the most limited amounts of project money for the commissions in fiscal 1969.

Why did this lack of commitment occur? Can it be traced to basic deficiencies in the nature of the planning process which was to be undertaken by the commissions? The classic and almost inevitable characteristic of most planning activities is that those who actually make the decisions which affect economic development are not involved in the creation of "rational" plans, preferring to retain their freedom of action. Planning thus becomes a process which is uninfluenced by, and does not influence, the real actors of change. However, the commissions were deliberately constituted so as to have continuous access to key decision makers in state houses and in the Federal government. Moreover the legislation was so formulated that really large regions, containing areas of rapid growth as well as areas of decline, could be designated. Similarly the legislation cannot be faulted on the commissions' authorized area of activity, which appeared to be enormous.[2] Nor can there be any doubt that the President promised to seek Federal financial support for projects which would be generated by the commissions.[3] But in the event why was this promise not honored?

There are two possible explanations. The first is that the President and his close advisers were unwilling to commit Federal resources to a large number of commissions, each one of which was submitting huge requests. In this view, efforts to help Appalachia were a one-time commitment occasioned by the profound impression of dire regional poverty made on President Kennedy during the West Virginia primary of 1960, the proven willingness of Appalachian state governors to work together, and a two and a half year preparation of remedial plans which Congress ultimately backed with hard cash. By contrast, it was assumed that the new commissions could not play such a crucial role, since they operated within less depressed areas which had shown no strong evidence of political cooperation across state boundaries prior to the possibilities of Federal assistance for regional planning.

Several facts lend corroboration to this interpretation. The President does not appear to have been in any hurry to make appointments to the positions of Federal cochairmen of the five commis-

2. See Appendix for the text of the act.
3. See Chapter 4, pp. 68–70.

sions. As Table 20 shows, there are three distinct stages to the formation of each commission—designation, appointment of cochairmen, and formal establishment—and this process was completed for all of the commissions only by the latter half of 1967, more than two years after the legislation had been passed.

It is also significant that the Federal cochairmen not only held the Federal Civil Service rank of assistant secretary, which was one position lower than that of the Appalachian Commission's cochairmen, but also reported to the Secretary of Commerce and not directly to the President, the route for the Appalachian appointee.

Table 20. Date of Establishing 5 Regional Commissions

Regional Commission	Designation	Appointment of Federal cochairman	Formal establishment
New England	March 1966	January 1967	March 1967
Upper Great Lakes	March 1966	August 1966	April 1967
Ozarks	March 1966	June 1966	September 1966
Four Corners	December 1966	August 1967	September 1967
Coastal Plains	December 1966	July 1967	July 1967

There is also some legitimate doubt as to the caliber of the Federal cochairmen in some of these commissions. Four out of the five appointments resulted from an initial selection by a congressional delegation, to which the President acceded, even though some of the cochairmen did not have extensive experience with the problems of generating economic development.

Two other acts of the President suggest a firm lack of commitment to the five new commissions. In 1967 the President sought congressional approval for separate funds to further extend the work of the Appalachian Commission. Congressional appropriation on almost exactly the scale sought by the President suggests strong grass-roots support for the efforts of the Commission and the President's recognition of the merits of continuing to give this agency separate budgetary funds.[4] However, when it came to supporting the projects of the other regional commissions, the President sought appropriations of only $10 million in fiscal 1969 and stressed that this money

4. "Appalachia teaches many lessons. Its recent progress tells us the partnership of Federal, State and local governments can work—in Appalachia and in other regions." Statement by President Lyndon Johnson, Dec. 29, 1967, on Executive Order 11386, "Federal Participation in Regional Economic Development."

should be administered under Title I of the 1965 act and thereby allocated with the explicit approval of EDA. The second fact is the President's obvious desire to maintain control over the work of the commissions and, in particular, his insistence that the Federal co-chairmen should be effectively integrated with the activities of Federal departments.[5] Both of these actions may be explained by the President's awareness that the commissions were likely to become the tools of the governors, many of whom were from the opposition party.

An entirely different interpretation is possible. It could be argued that the establishment of the commissions was inevitably a time-consuming process. The problems of building research teams, of designating areas, and above all of creating a distinctive role for each commission in relation both to existing state planning agencies and to individual Federal agencies inevitably meant that no immediate solutions were likely to be found. Thus each new institution was being challenged to find solutions which were appealing to individual states, to Federal agencies, and to congressmen and senators from the affected states, and such a challenge could not receive a realistic response in any short period of time. This view is related to another. The increased commitment to the Vietnam War occurred at precisely the period when several commissions were actually bringing forward their detailed proposals for assistance and detailed strategies for development. In common with many other domestic programs, the regional approach was clearly seen as expendable, at least in the short run. Whether this implied presidential disapproval over the longer run therefore cannot be deduced from the available evidence.

In a very real sense, which of these interpretations of events between 1965 and 1968 more accurately fits the facts is unimportant. A period as short as this cannot be interpreted as proof of either the validity or the invalidity of the regional approach. Such an evaluation requires us to stand back from the pressure of political events to establish why the Federal government should be interested in a regional approach.

5. Executive Order 11386, Section 1(C), "The Secretary of Commerce shall . . . provide guidance and policy direction to the Federal Co-Chairmen—Review the regional economic development plans and programs submitted to him by the Federal Co-Chairmen, budgetary recommendations, the standards for development underlying those plans, programs and budgetary recommendations, and legislative recommendations; and advise the Federal Co-Chairmen of the Federal policy with respect to those matters and, where appropriate, submit recommendations to the Director of the Bureau of the Budget."

Some Legitimate Arguments for Federal Involvement

The most obvious reason for a conscious regional approach is the fact that the Federal government, through its complex of activities, does have a profound effect on the level and content of economic activity in every state in the Union. In 1965, for example, the Federal government expended 34 percent of the funds for general domestic programs,[6] and in some categories of expenditure the Federal share was dominant or very important (Table 21).

Table 21. Expenditure in Selected Domestic Programs, 1965

(In millions of dollars)

Category	Level of government			
	Total [a]	Federal	State	Local
Postal service	5,058	5,058		
Education	30,021	2,727	14,532	22,810
Highways	12,348	4,124	4,844	4,039
Natural resources	10,990	9,447	1,250	519
Health and hospitals	7,671	2,601	2,942	2,741
Housing and urban renewal	2,198	1,625	80	1,227
Air transportation	1,198	853	64	360

[a] Totals exclude duplicative transactions between levels of government.

Many forms of Federal activity, though established for reasons of national interest, do have a discriminatory locational impact. Tariff policy protects specific industries in specific regions, defense expenditure favors certain regions more than others, space program outlays have resulted in entirely new urban areas being established, agricultural subsidies maintain incomes in rural regions, oil depletion allowances affect factor returns in Gulf states, and so on. In this context, it is worth underlining the fact that EDA expenditures, which averaged $300 million annually for the whole country, were dwarfed by the outlays of several other agencies. Table 22, which summarizes the regional expenditures of some of these other agencies, shows that the Departments of the Interior, Transportation, Army, and Agriculture, the Office of Economic Opportunity, the Small Business Administration, and probably the Department of Labor all spent more on regional development than EDA.

6. Direct general expenditures excluding those for defense, space, international programs, trust funds, and government-operated enterprises; *Regional Economic Development*, Oct. 1967, Part I, p. III–18.

More generally the impact of Federal grant-in-aid programs is increasing rapidly. In 1946, the Federal government spent $894 million to help state and local governments augment their public programs. By 1960 this amount had climbed to $7.04 billion, and in 1967 it reached an estimated $15.366 billion.[7] At this last date, there were 170 separate Federal aid programs which required 150 bureaus in Washington and 400 field offices to administer the programs. Even more significant from the point of view of regional planning is that 100 of these programs stipulated, as a precondition of receiving Federal assistance, that subnational planning should precede and accompany any expenditure flows. Indeed, in 29 of these programs, planning for the development of specific areas was the goal, and in many programs the overall objective was specifically stated to be economic development. For example, 25 programs were concerned with job creation, 40 sought the economic sustenance of farmers and their families, 44 provided technical and financial assistance to industry and commerce, and 52 aided state and local governments, nonprofit organizations, and small businesses.[8]

The grant-in-aid system is particularly suited to U.S. conditions, since local agents define the content of a problem and are left to administer remedial strategies while the central government has the responsibility for determining the importance to the nation of solving a given subnational problem, of delineating solutions, and of judging whether nationally raised funds have been spent wisely. Thus the initiative rests with those who are most likely to understand the intricate factors involved in the successful local implementation of a given program, while the national government is freed from establishing an enormous bureaucracy each time a new program is introduced.

And yet, this system can create, rather than solve, administrative problems. Extensive local participation may bring into being powerful pressure groups which effectively bar a rationalization of programs even though all the evidence points to the need for dropping some programs in favor of new approaches. In consequence new programs tend to be added without old ones disappearing, old objectives mingle with new ones, and old criteria of eligibility for funds complicate the search for new criteria.

The solution to these problems appears simple, the scope for coordination enormous. It would, however, be naive to suppose that

7. *Regional Economic Development,* Oct. 1967, Part I, p. III–16.
8. *Regional Economic Development,* Oct. 1967, Part I, p. III–22.

Table 22. Indicative Summary of Selected Agency Expenditures by Regional Designations for FY 1967 or FY 1968 or FY 1969

(In thousands of dollars)

Regional unit	Executive agency									Federal dollar outlays[k]
	Interior[a]	Transportation	Army[e]	HUD[f]	HEW	Agriculture[h]	Labor[i]	OEO	SBA[j]	
Alaska	$51,169	[b]$40 Hwy	$205,742		[g]$50,629	$7,757 FS 132,800 REA	$3,677	$3,000	$62,886	$672 mil.
Appalachia	73,635	807 Hwy [c]1,300 UMT [d]9,502 FAA	2,000,000		34,000	27,780 FS 222,000 SCS 900,000 REA	78,865	103,000	51,180	10,659 mil.
Coastal Plains	18,170	151 Hwy 131 FAA	80,520		457	3,900 FS 299,000 SCS 360,000 REA	21,222	27,000	19,645	4,147 mil.
Four Corners	176,902	217 Hwy 155 UMT	42,974		550	319,000 REA 98,000 SCS 45,000 FS	10,369	38,000	14,579	2,040 mil.
New England	54,228	385 Hwy 35,145 UMT 220 FAA	405,455		655	5,930 FS 6,000 SCS 46,000 REA	56,741	54,000	53,050	10,395 mil.

Ozarks	27,421	86 Hwy	111,431	......	11,850 FS	15,152	22,000	12,132	2,033 mil.
		44 FAA			188,000 SCS				
					630,000 REA				
Upper Great Lakes	20,100	278 Hwy	27,656	......	24,300 FS	14,562	25,000	10,919	1,564 mil.
		262 FAA			85,000 SCS				
					490,000 REA				
									$31,510 mil.

Notes:
From U.S. Department of Commerce, Federal Advisory Council on Regional Economic Development (initial meeting Oct. 22, 1968), Dec. 1968, App. F, p. 260.

Fiscal year data vary as indicated and in some few cases involve a cumulative total over several years. Accordingly, they are not entirely additive.

[a] FY 1969 data.

[b] Calendar year 1967 data (highway capital outlays).

[c] FY 1968 data (urban mass transit).

[d] Cumulative federal aid to airports program.

[e] Corps of Engineers program including only preauthorization studies, authorized (but unfunded) projects, and projects under design.

[f] Figures for HUD cannot be broken down on this regional basis, but major programs for FY 1969 include $724 million – model cities; $750 million – urban renewal; $165 million – water and sewer.

[g] FY 1967 – other data FY 1968. (Appalachia figure is for FY 1969.)

[h] Consists of FY 1968 Forest Service data and cumulative REA loans.

[i] FY 1968 data on manpower work and training programs.

[j] Loan activity for FY 1968 – only SBA share of loans indicated: total *value* of loans runs about 20% greater.

[k] *Not* a total from this table: rather the governmental total produced from the federal outlay data series of OEO.

efforts to rationalize and coordinate existing programs at the Federal level could generate quick results. Since there is no agreed set of national goals, and no supreme agency to enforce administrative action, schemes of rationalization are likely to be time consuming, frustrating, and only successful where individual agencies perceive obvious benefits in cooperating with other agencies. Nevertheless, there is an overwhelming case for some agency to attempt to avoid *future* inconsistencies and conflicts in any proposed legislation and major Federal action which explicitly deals with economic development at the subnational level. Here the objectives would be to diagnose current and future regional performance, evaluate the impact of proposed legislation and Federal measures on subnational areas, delineate blatant inconsistencies in separate Federal economic development programs, and perhaps actually measure the impact of actions which have been taken.

This type of activity would require an office in Washington, a high-ranking official in each major Federal department, and some system of regional offices. Above all, it requires presidential recognition that questions of "why," "what," "for whom," and "for how long" are inseparable from questions of "where" when Federal action is being considered and evaluated. Thus strong presidential pressure would have to be exerted to ensure that interdepartmental conflicts were openly discussed and settled, and one department would, almost inevitably, have to be singled out as the initiator of cooperative activity.[9]

Is this as far as the Federal government should go with a policy for regional economic development? Is regional surveillance, diagnosis, and forecasting as an aid to better Federal program formulation and implementation an adequate goal? Should the institutional machinery simply attempt to coordinate the actions of the executive in response to the myriad requests from individual regions and areas?

There are numerous aspects to any attempted answer to these questions, but four broad factors must be considered. First, what is the nature and importance of the economic benefits which the articulation and implementation of a Federal regional economic policy would generate? Secondly, what kind of planning is involved? Is it planning with precise numerical targets at some specified terminal

9. Many of these institutional elements were present in President Johnson's proposed Regional Economic Development Advisory Council; *Regional Economic Development*, Dec. 1967.

date, and a clearly defined sequence of measures established as the means of reaching the desired end state? Alternatively, is the objective of the program specified in vague terms, and the policy to be taken by the national government mentioned but not given priority weighting on an explicit sequence for implementation so that the overall concept of planning is simply to indicate the possible future course of government activities? Thirdly, what spatial focus does the planning have? Is it primarily aimed at spatial pockets such as individual cities, counties, or labor market areas (intraregional planning), at individual regions (regional planning), or at the totality of regions which constitute the nation state (national spatial planning)? Fourth, in what ways do subnational levels of government participate in the formulation of program goals and the implementation of commonly agreed on policies?

The allegation of economic benefits flowing from an articulated regional economic policy is based on the premise that the free movement of goods, capital, people, and ideas necessarily creates a nation state of interdependent regions, but not necessarily an optimal spatial distribution of economic activity and population. Suboptimality may occur for three distinct reasons. First, a relatively low overall rate of national unemployment may mask an extremely low level of unemployment in certain regions and persistently above average unemployment in other regions. A nationally stimulated increase in the demand for labor in the labor-surplus regions may move the whole economy closer to the margin of full employment without causing inflationary pressures. Secondly, the rapid growth of certain regions, and especially the metropolitan areas within these regions, and the rapid decline of other regions do not necessarily represent an optimal development path. Investors in the fast-growing metropolitan areas may underestimate the cost advantages of operating in depressed areas, may by their actions create a net demand for public infrastructure investment while spare public overhead capacity exists in the depressed regions, and may cause unwanted externalities precisely because the social costs of their actions cease to become individually identifiable as the metropolitan area grows in scale. Thus the fast-growing region may expand past the point where the total public and private net capital-output ratio is minimized. In contrast lagging regions, which present real opportunities for a lower net capital-output ratio, may nevertheless continue to decline. This may occur because regional agents are unskilled in the identification and eradication of structural weaknesses and because the migrants who quit the region are subsidized by receiving communities which fail to capture the marginal costs of community services consumed by the migrants because of lags in tax payments.

Even if there is no connection between the growth of particular regions and the decline of other regions the third argument is that there is a case for intervention simply because the growth of population and of economic activity does not occur in the right places and on the right scale. Here, the assumption is that major cost savings are associated with accommodating population growth in large planned communities, where infrastructure can be planned on a large scale and full utilization can be guaranteed. One final justification is relevant, though it relates to equity rather than efficiency criteria, and that is the equalization of per capita income by region.

We have already reviewed the validity of these arguments in Chapters 2 and 3, and there is no need to repeat the discussion. We can, however, delineate some critical implications for Federal regional policy.

1. The sixties have witnessed a fast rate of employment growth in every region of the country and per capita income convergence towards the national figure in those states which have had persistently lower than average incomes in the past. At the root of these major tendencies is a growing urbanization of the U.S. population and, in particular, an above-average growth rate for urban areas of more than 50,000 population.

2. Despite the extremely fast growth of the economy since 1960, imbalances between the demand for labor and its supply are very much in evidence in particular regions and within some major metropolitan labor markets. Moreover, as the growth rate of the economy slows down there is a strong possibility of a growing dispersion between the highest unemployment rates and the average rate of unemployment.

The cost of a mismatch between the demand for labor and its supply can be measured in output foregone. But in addition, when an economy is operating close to the margin of full employment, a relatively inelastic supply curve may cause price increases which affect the export strength of the economy. Therefore as long as the market does not generate a satisfactory match between the demand for labor and its supply, and as long as these imbalances are apparent in several regions of the country, there are legitimate grounds for a Federal regional policy to achieve spatial balance.

3. The dangers of misallocation of resources by local agents seeking the widespread revitalization of depressed areas are real. The economic claims on behalf of many such depressed areas are spurious, and the case for a use of Federal resources to discern and initiate viable redevelopment solutions is strong. Such solutions may include aid for area renewals, encouragement to labor migration, and the deliberate build-up of existing regional urban centers.

4. The argument that the growth of large metropolitan areas should be deliberately restrained because they suck the economic lifeblood (i.e., the high-quality labor) from less buoyant regions has been greatly exaggerated. Moreover, the dangers of undue polarization are slight in a country which has spawned several major metropolitan areas in this century and has many cities of more than one million population.

5. The argument in favor of diverting population growth from large cities to smaller centers on the ground that major cities currently are high-cost locations (i.e., high private costs and high social costs) is unproved and cannot be used to justify an explicit national policy on population redistribution. In any event even if redistribution is worth while it is largely an intrastate problem.

6. The argument that population growth in planned centers would be cheaper than population growth spread throughout growing cities of all sizes may have some validity, but the cost savings are probably of a small magnitude.

All of this suggests that Federal regional policy should be concerned primarily with facilitating intraregional and intrametropolitan labor market efficiency and with protecting *individuals* who are incapable of adjusting to competitive changes rather than with encouraging parametric shifts in the distribution of population and of economic activity. As a general principle, regions and metropolitan areas should be treated like private companies, that is as competitors for mobile national resources of labor, capital, and ideas. Thus, whereas it is legitimate for the Federal government to encourage a high level of aggregate demand, factor substitution, factor mobility, and the control of unwanted externalities within the economy at large, it is not legitimate for it to advocate one spatial pattern of economic activity and population nor to bias private rates of return, as for example by offering tax concessions or cheap overhead capital, in order to achieve this supposedly optimal distribution.

In effect, therefore, the Federal government should not establish and try to achieve targets which break down future national output, investment, consumption, and saving by regions, or even create regional targets for population and interregional migration. However, it should be explicit on where, in the future, it expects market forces to be insufficient to move people out of poverty or insufficient to prevent a serious underutilization of local labor talents, and where it feels immediate Federal intervention would be effective. A dual role of advance area surveillance and diagnosis of remedy on a *total, all region* basis should be the core of the Federal regional approach.

But where do the regional commissions fit into all this? If the Federal government does not require the commissions to implement an explicit population/activity pattern and if rationalization of future Federal programs which have a marked spatial impact proceeds from Washington downward and not from the localities to Washington, is there any validity in a multistate approach? This question must be further refined. Should the Federal government continue to provide project support for the commissions such as it has provided for the Appalachian Commission or should its objective be to sustain purely advisory multistate agencies?

There are numerous reasons why the Federal government should think twice about funding projects conjured up by the commissions. The most obvious is the scope for political dissension among the principal "actors" in the multistate planning process. The system requires that president, senators and representatives in Congress, governors, and state politicians must be in accord on basic strategies even though there is a very high probability that more than one party will be represented in the overall process. For example, it will require acts of political magnanimity for a president to allocate funds for regions in which opposition party governors are in the majority. It may also be asking too much of any individual governor to join in decision making which may result in his state's receiving less investment than associated states. This may present especially acute problems when distant and seemingly more prosperous states are allocated an "unfair" share of the overall investment. Given the huge territorial size of the designated regions and the likelihood that each state will vary in its technical ability to make a case for a more than proportionate investment, then such an outcome is always possible. Moreover this process destroys a canon of democratic theory, namely, the explicit identification of political responsibility for specific acts of policy. It is of course always true that ruling parties within each state will exaggerate the extent to which they alone have been responsible for political successes and in contrast will hide behind excuses about external interference when their policies are unsuccessful. In any case, the regional commission system of multiple decision makers makes it decidedly easy for state governors to explain away their misutilization of Federal funds by stressing that they were obliged to accept majority decisions even though they had not personally favored the unsuccessful policies.

Another criticism could be leveled at this approach. Appalachia is often cited as an example of a region which is now better planned because of its commission. It is probably true, however, that the Appalachian Commission has been singularly successful because its major activity has been the planning of the interstate and intra-regional highway program, which clearly benefited every state and obviously required multistate participation over the alignment of the roads. Other types of planning activity, for example, the selection of regional growth centers, may not present such obvious opportunities for mutually beneficial decision making.

These disadvantages suggest that the Federal government should not seek to channel its funds through this type of institution. Nevertheless, there are the obvious dangers in the current system, where individual localities may approach individual Federal agencies, that projects may be funded without a comprehensive awareness of the likely costs and benefits of the funding or indeed even of the possibility that directly conflicting policies may be followed by individual agencies.

Secondly, there are types of planning activity which must be considered on a multistate level. For example, if one state is considering a policy of establishing new towns or of encouraging rural out-migration, such activities may affect service hinterlands within neighboring states and the alignment of communication lines. For both reasons, there is a case for advisory multisate commissions which would advise state planners on redevelopment strategies, act as a sieve for major projects emanating from within individual states and requiring Federal backing, advise individual Federal agencies on local development problems and opportunities, and perhaps run training programs for planners within individual states. Of course there is no reason why such institutions should not encourage individual states to join with each other in approaching Federal agencies for major project help. Similarly, it may be reasonable for the commissions to act as paid consultants on specific development problems cited by individual states.

All this leads to the conclusion that there is a useful role for multi-state planning commissions, provided they are basically viewed as advisory units serving the individual states with problems of concentrated spatial distress, and for Federal agencies keen to alleviate such distress.

7

CONCLUSIONS

Those who advocate a major role for the Federal government in the shaping of regional economic development must satisfy their critics (and themselves) on several separate scores. They must first establish why the market mechanism cannot be relied upon to generate an optimal spatial distribution of population and of economic activity.

Given current economic conditions it seems highly unlikely that such a conclusion could be reached, for it is clear that the prodigious growth of the economy in the second half of the sixties not only guaranteed national full employment but also generated rapid income and employment growth in every region of the country. And yet a more intensive appraisal establishes beyond doubt that rapid national growth has not eradicated severe and persistent imbalances between the demand for labor and its supply in literally hundreds of small and medium-sized labor markets spread throughout every region of the Union.

Estimates of the seriousness of this problem vary, but we have shown that if unemployment were reduced to 6 percent in all these problem areas, then 130,000 workers would be added to the employed labor force and the economy would benefit to the extent of a $1.3 billion rise in national output. In addition, in those areas in which extremely low incomes and massive underemployment coincide, 150,000 families would be assured a politically acceptable income if Federal targets were reached.

There is enormous variety in the initiating causes of labor market disequilibrium, but what is often apparent is the speed and ferocity of the forces which denude the local economic base. A drastic fall in the demand for a major local export good, the rapid substitution of capital for labor in a key local sector, the outmigration of local manufacturing activities, the closure of a military base, an excessive inmigration of low-productivity workers, these and many other events may presage severe unemployment, underemployment, and wage differentials for given skills.

Now in such conditions market processes do work and some labor and some capital are redistributed to more productive activities outside the local area. For many affected areas, however, the unsubsidized outmigration of labor is unlikely to be a sufficiently elastic response to the decline in employment opportunities. Moreover, there are real possibilities that the qualitative differences between outmigrants and inmigrants will denude the stock of labor skills and that local governments will be forced to downgrade educational programs, leaving the nation to bear the long-run cost of a future supply of low-quality workers.

Given these possibilities, there are four basic reasons why the allocation of *short-term* subsidies to labor-surplus areas may benefit the nation. With the economy at the margin of full employment, aggregate demand increases outpacing increments to supply, and factor prices rising sharply, pockets of unemployed, underemployed, and badly educated employees constitute a pool of labor which when fully integrated into the employed labor supply could help to dampen the inflationary consequences of demand increase. Thus, where surplus labor in the lagging areas cannot be induced to move to areas of greater opportunity, then subsidies which result in the creation of local work opportunities will generate a larger increase in national output than if such subsidies were used in the full-employment regions. Secondly, rapid unemployment decline may leave areas stranded with utilizable but under-used social and economic overhead capital. Even though the private rate of return on capital in such areas is relatively low, the very small marginal public investment cost of accommodating private growth may justify granting subsidies to encourage such private development. Thirdly, the impoverishment of unemployed but immobile older workers who are dependent for their income upon the depleted revenues of local units of government may warrant intervention in the form of supplementary unemployment or early retirement benefits. Finally, there is an obvious case for using subsidies to compensate for inadequate local expenditures on education.

If these are legitimate reasons for aiding distressed areas to overcome their short-run frictional problems, the second crucial question is what should be the long-run objective of subsidization. There are two aspects to this question. The first is whether distressed areas should have their level of demand for labor artificially increased so that migrants do not flood to metropolitan areas under the lure of higher welfare payments or low initial taxation, where they would simply exacerbate existing diseconomies of scale. This type of argument is seriously deficient, since it takes no account of the productivity gains which accrue to the nation when workers transfer from low-output to higher-output activities. In any event, the recent evidence suggests that the main causes of population growth in metropolitan areas are natural population increases, international migration, and net inflows of people from other large labor markets. The second part of the question is even more important. Can subsidies be advanced over the long run on the assumption that every distressed area can learn how to redevelop provided it is given help over the critical stages of "structural shock"?

We have concluded that such a view is naive since market forces may be so powerful as to destroy the economic rationale of many local economies. As a general rule the relatively small and very small labor markets of under 50,000 population are likely to face the test of changing patterns of demand and of competition with the least resilience. More specifically we have categorized distressed areas into five broad types, that is, rich labor-surplus areas in which massive inmigration has caused disequilibrium, old industrial and urbanized areas with ossified export structures, relatively productive rural areas in which heavy unemployment has been caused by capital substitution, and the really poor, unproductive rural and worked-out mining areas which have very low per capita incomes, heavy outmigration, and in some instances heavy unemployment as well.

A combination of market trends, the size of the affected areas, and the positive internal response to disequilibrium suggest that many areas within the first three types of area may, once again, experience a reasonable rate of per capita income growth albeit with a diminished population or a retardation in the rate of population growth. But for the vast majority of the really poor rural and mountain areas, and especially those which are not within immediate travel-to-work reach of a major urban area, the chances of revival are negligible. For such areas, solutions which prop up the demand for local labor over the long run are likely to be inordinately wasteful.

How does the Federal government fit into all this? If we accept that there are legitimate reasons for short-term subsidization to overcome market imperfections and that there are reasonable prospects for long-term redevelopment in many currently distressed areas, though not the small, really poor communities, what justification is there for national intervention? Put in another way, can the work of redevelopment and the subsidies which accompany it be placed entirely in the hands of individual states?

There are four sound reasons why the Federal government must actively participate in the redevelopment process. First, the Federal government is in a unique position to *advise* on methods of achieving successful local redevelopment partly because of its international connections, and especially its aid-giving activities, and partly because of its concern for national development as distinct from more narrowly localized goals. Secondly, as a result of its massive expenditures, and particularly its defense outlays, the Federal government cannot avoid differentially affecting the rate of economic growth of the several regions of the country. There is, in consequence, a genuine need for sophisticated *surveillance* and *forecasting* of the multiple impacts of Federal activities and how these may initiate and maintain localized disequilibrium. Thirdly, the competitive actions of individual states and localities which are seeking to lure exogenous investment and enterprise to their problem economies may enhance the profitability of mobile industrial activities without securing any permanent remedial effect on the problem economies. Thus the offer of Federal subsidies to any given lagging areas may create a *framework of information* on the actions of competitor lagging areas which results in a more rational set of local subsidy practices. Finally, and this is the really important argument, the *opportunity cost* of not being actively involved with the redevelopment of local areas may be that some states and many smaller areas will fail to exploit their opportunities successfully.

It is precisely at this point, however, that Federal regional economic policy is most difficult to formulate. Over any given period, the financial resources available to those Federal agencies which are directly concerned with "place prosperity," as distinct from the prosperity of given human groups, industries, or natural resources, are likely to be severely limited. Moreover, the normal political processes are likely to generate a distressed area set which is so numerous that Federal project funds are seemingly totally inadequate.

This imbalance between apparent opportunities and actual Federal resources for redevelopment was at the root of difficulties faced by EDA. With an annual average of $300 million to spend and approximately 900–1000 areas eligible for assistance, some system of priorities had to be devised. Several approaches were tried. The most popular politically was to use the project "queue" as the means of establishing priorities. Thus projects were funded in any distressed area which had submitted the necessary evidence to suggest that there were some possibilities of a permanent rise in local employment as a result of the project.

Though this approach satisfies the political criterion that many areas should be seen to receive an equal chance of Federal help, it is likely to be highly unsuccessful as a method of economic redevelopment. Most notably it puts the funding agency into the role of constantly reacting to project submissions, oftentimes from areas with extremely limited planning capabilities, and tends to encourage a bureaucratic system in which the different types of assistance are handled by separate parts of the agency. The net outcome is that the agency can mass its projects to a very limited extent and tends to be constantly moving on to the next area, without any real possibility of a long-term commitment for any given area.

Worries of this kind resulted in a courageous attempt to establish concrete area priorities for EDA investment. The "worst-first" policy was consistent in its objectives, easy to understand, and obviously concerned with long-term developmental pay-offs rather than short-term political popularity. However, there is a strong danger that under such an approach employment benefits are restricted to the period of construction of infrastructure. Thereafter, the likelihood in many cases is that local communities will bear a heavy debt burden as the new infrastructure fails to attract the expected increase in economic activity.

There is a strong case for preserving some elements of the worst-first approach, but in a recast form. We have already argued that there is enormous variation in the possibilities of redeveloping areas within the current distressed area set. With many, though not necessarily all, of the areas classified as rapidly growing and rich, old industrial, and not-so-poor rural, the growth of the economy as a whole, together with the efforts of individual states, should result in successful redevelopment. Federal assistance to these areas should be strictly limited, both in quantity and in kind, and serve to encourage the individual state to take necessary remedial action. In terms of the possible reasons for Federal intervention discussed previously, EDA assistance should consist of advice on redevelopment strategies, surveillance and forecasting of the impacts of current and future Federal activities on given areas, and information on the activities of competitive lagging areas. In some instances, this range of assistance should be extended to include direct assistance to state planning agencies which have shown a willingness, but do not have the resources, to devise redevelopment strategies. This means that the bulk of EDA activity and resources should be directed to solving the economic problems of some of the most seriously depressed local economies in the persistent and substantial unemployment category, the really low income counties, and the few areas with very low incomes and serious unemployment.

There can be no one remedial solution for these areas, but we have argued that there is a strong case for concentrating EDA resources in a few relatively populous centers which are accessible to the distressed areas but have not experienced persistent structural decline. Here the ultimate objective is to lure people away from depressed localities which have extremely limited chances of redevelopment to areas which are deliberately chosen for subsidized, rapid growth.

There are many reasons why such an approach makes sense from a national viewpoint. Spatially concentrated investment is likely to have a relatively large regional income multiplier effect in the short run, to divert more economic activity from other areas at a given subsidy cost, and, in adding to existing agglomeration economies, to minimize private and public costs of production. In addition, any encouragement which is given to the concentration of regional population within a few relatively large population centers is likely not only to minimize the per capita costs of providing social and economic overhead capital but also to fully exploit scarce planning skills. Moreover, indicative plans for public investment can be given a rigorous spatial dimension, and a real effort can be made to coordinate the activities of individual Federal agencies from below. However, the critical argument in favor of this approach is that where some residents of lagging communities will not uproot themselves and move long distances to a nonregional center, they may well migrate to the local center. The availability of work with a limited skill content may act as the main attraction, but the resource cost of migration may also be small, and there are obvious possibilities of maintaining close ties with family and social groups in the original environment. At best the regional center may generate sufficient new job opportunities to attract and hold the migrant over the long term; at worst it should provide a "staging post" where the migrants' learning costs of urban assimilation are low and an upgrading of migrant skill is possible.

EDA's experience with creating and establishing development centers has been too short lived to permit us to evaluate whether the strategy was well conceived. Moreover, there can be little doubt that the development center approach was accorded a somewhat uncertain priority during the internal discussion on the worst-first strategy. Two facts are clear, however. EDA's assistance should be far more broad gauge. The subsidized expansion of development centers is a process which requires a whole range of inducements and should include direct subsidies to incoming and local companies, labor market information to would-be migrants within the region, moving and installation grants, retraining subsidies, and perhaps housing subsidies. Secondly, more than just lip service must be paid to the problems of integrating the activities of the different Federal agencies which, by their activities, have an impact on the growth of the development centers.

Inevitably, essential to adherence to a development center policy is a belief in results over the long run. Local rivalries must be reduced, the benefits of investment concentration for the whole lagging area must be delineated, and specific goals and program strategies must be devised. Not least, planning activities must be allied with the local structures of political power. All of these are problems internalized by the operational requirements of a development center approach. But there are several crucial questions which cannot be answered at this territorial level. In any given lagging region, how many development centers should there be? Should Federal regional policy encourage active competition for resources among development centers or try to delineate noncompeting hinterlands in which an individual center provides the dynamic focus for a territorially integrated labor market? How should urban centers not chosen for discriminatory assistance be aided with their adjustment problems? What type of administrative machinery should exist to initiate and coordinate public investment strategies for the lagging region as a whole?

The answers given in the legislation of 1965 seem to suggest that the long-term solution to these problems is to be found in the activities of Regional Action Planning Commissions which contain Federal representatives, state political representatives, and skilled technical planners. Insofar as the expertise and local knowledge of such commissions are expected to help rationalize the flow of Federal funds to any region, whether as part of a specific type of investment, specific agency program, or general assistance for economic development, then there may be merits in multistate machinery. But to expect such commissions to initiate and coordinate Federal and state efforts in regional development and to seek common agreement for specific regional investment strategies is naive and misleading. In part this results from using a prototype of U.S. regional planning machinery which does not have general applicability. The Appalachian Regional Commission was nurtured in conditions of high national unemployment, spatially concentrated socio-economic distress of staggering magnitude, a long-standing series of contact among governors in the affected states, genuine presidential interest, and a careful and prior appraisal of development potential and the resources required for remedial action. With the five new commissions none of these conditions applied in the years between 1965 and 1968. As a result, the Federal government has almost entirely ignored the commissions' requests for project finance.

The crucial point is that successful development planning must somehow integrate the preferred development strategies of technical planners with the preferences of politicians charged with responsibility for allocating resources for the redevelopment of lagging economies. For the foreseeable future this fusion between political commitment, political responsibility, and technical guidance is likely to be achieved only at the state level. At most, then, any multistate institutions should be concerned with the training of state planners, the initiation of multistate cooperative developments, and the provision of regional forecasts and technical advice.

Thus Federal regional policy should rest securely between a commitment to highly localized relief and to lagging region revival, on the assumptions that polarization of economic activity will continue and should be encouraged and that Federal assistance must be channeled through subnational political agencies which ultimately are answerable to a specific electorate for the quality of their investment decisions.

Therefore the most encouraging feature of EDA activity is that the temporary dalliance with allocation based on a wide scattering of project money and the later concentration upon helping areas with extremely limited potential for economic development have increasingly been replaced by a commitment to a development center approach. This is a signal achievement which simultaneously destroys the myth that development potential is distributed equally over space and creates a real possibility for a Federal strategy to fulfill the oft-repeated promises of rational economic development for lagging areas.

APPENDIX

PUBLIC WORKS AND ECONOMIC DEVELOPMENT ACT OF 1965

An Act

To provide grants for public works and development facilities, other financial assistance and the planning and coordination needed to alleviate conditions of substantial and persistent unemployment and underemployment in economically distressed areas and regions.

Be it enacted by the Senate and House of Representatives of the United States of America in Congress assembled, That this Act may be cited as the "Public Works and Economic Development Act of 1965".

Public Works and Economic Development Act of 1965.

STATEMENT OF PURPOSE

SEC. 2. The Congress declares that the maintenance of the national economy at a high level is vital to the best interests of the United States, but that some of our regions, counties, and communities are suffering substantial and persistent unemployment and underemployment; that such unemployment and underemployment cause hardship to many individuals and their families, and waste invaluable human resources; that to overcome this problem the Federal Government, in cooperation with the States, should help areas and regions of substantial and persistent unemployment and underemployment to take effective steps in planning and financing their public works and economic development; that Federal financial assistance, including grants for public works and development facilities to communities, industries, enterprises, and individuals in areas needing development should enable such areas to help themselves achieve lasting improvement and enhance the domestic prosperity by the establishment of stable and diversified local economies and improved local conditions, provided that such assistance is preceded by and consistent with sound, long-range economic planning; and that under the provisions of this Act new employment opportunities should be created by developing and expanding new and existing public works and other facilities and resources rather than by merely transferring jobs from one area of the United States to another.

TITLE I—GRANTS FOR PUBLIC WORKS AND DEVELOPMENT FACILITIES

SEC. 101. (a) Upon the application of any State, or political subdivision thereof, Indian tribe, or private or public nonprofit organization or association representing any redevelopment area or part thereof, the Secretary of Commerce (hereinafter referred to as the Secretary) is authorized—

(1) to make direct grants for the acquisition or development of land and improvements for public works, public service, or development facility usage, and the acquisition, construction, rehabilitation, alteration, expansion, or improvement of such facilities, including related machinery and equipment, within a redevelopment area, if he finds that—

(A) the project for which financial assistance is sought will directly or indirectly (i) tend to improve the opportunities, in the area where such project is or will be located, for the successful establishment or expansion of industrial or commercial plants or facilities, (ii) otherwise assist in the creation of additional long-term employment opportunities for such area, or (iii) primarily benefit the long-term unemployed and members of low-income families or otherwise

52-756—(172) O-65

substantially further the objectives of the Economic Opportunity Act of 1964;

78 Stat. 508.
42 USC 2701 note.

(B) the project for which a grant is requested will fulfill a pressing need of the area, or part thereof, in which it is, or will be, located; and

(C) the area for which a project is to be undertaken has an approved overall economic development program as provided in section 202(b)(10) and such project is consistent with such program;

(2) to make supplementary grants in order to enable the States and other entities within redevelopment areas to take maximum advantage of designated Federal grant-in-aid programs (as hereinafter defined), direct grants-in-aid authorized under this section, and Federal grant-in-aid programs authorized by the Watershed Protection and Flood Prevention Act (68 Stat. 666, as amended), and the eleven watersheds authorized by the Flood Control Act of December 22, 1944, as amended and supplemented (58 Stat. 887), for which they are eligible but for which, because of their economic situation, they cannot supply the required matching share.

16 USC 1001 note.

(b) Subject to subsection (c) hereof, the amount of any direct grant under this section for any project shall not exceed 50 per centum of the cost of such project.

(c) The amount of any supplementary grant under this section for any project shall not exceed the applicable percentage established by regulations promulgated by the Secretary, but in no event shall the non-Federal share of the aggregate cost of any such project (including assumptions of debt) be less than 20 per centum of such cost. Supplementary grants shall be made by the Secretary, in accordance with such regulations as he shall prescribe, by increasing the amounts of direct grants authorized under this section or by the payment of funds appropriated under this Act to the heads of the departments, agencies, and instrumentalities of the Federal Government responsible for the administration of the applicable Federal programs. Notwithstanding any requirement as to the amount or sources of non-Federal funds that may otherwise be applicable to the Federal program involved. funds provided under this subsection shall be used for the sole purpose of increasing the Federal contribution to specific projects in redevelopment areas under such programs above the fixed maximum portion of the cost of such project otherwise authorized by the applicable law. The term "designated Federal grant-in-aid programs," as used in this subsection, means such existing or future Federal grant-in-aid programs assisting in the construction or equipping of facilities as the Secretary may, in furtherance of the purposes of this Act, designate as eligible for allocation of funds under this section. In determining the amount of any supplementary grant available to any project under this section, the Secretary shall take into consideration the relative needs of the area, the nature of the project to be assisted, and the amount of such fair user charges or other revenues as the project may reasonably be expected to generate in excess of those which would amortize the local share of initial costs and provide for its successful operation and maintenance (including depreciation).

(d) The Secretary shall prescribe rules, regulations, and procedures to carry out this section which will assure that adequate consideration is given to the relative needs of eligible areas. In prescribing such rules, regulations, and procedures the Secretary shall consider among other relevant factors (1) the severity of the rates of unemployment in the eligible areas and the duration of such unemployment and (2) the income levels of families and the extent of underemployment in eligible areas.

(e) Except for projects specifically authorized by Congress, no financial assistance shall be extended under this section with respect to any public service or development facility which would compete with an existing privately owned public utility rendering a service to the public at rates or charges subject to regulation by a State or Federal regulatory body, unless the State or Federal regulatory body determines that in the area to be served by the facility for which the financial assistance is to be extended there is a need for an increase in such service (taking into consideration reasonably foreseeable future needs) which the existing public utility is not able to meet through its existing facilities or through an expansion which it agrees to undertake.

(f) The Secretary shall prescribe regulations which will assure that appropriate local governmental authorities have been given a reasonable opportunity to review and comment upon proposed projects under this section.

SEC. 102. (a) In addition to the assistance otherwise authorized, the Secretary is authorized to make grants in accordance with the provisions of this title to those areas which the Secretary of Labor determines, on the basis of average annual available unemployment statistics, were areas of substantial unemployment during the preceding calendar year.

(b) Areas designated under the authority of this section shall be subject to an annual review of eligibility in accordance with section 402, and to all of the rules, regulations, and procedures applicable to redevelopment areas except as the Secretary may otherwise prescribe by regulation.

SEC. 103. Not more than 15 per centum of the appropriations made pursuant to this title may be expended in any one State.

SEC. 104. No part of any appropriations made pursuant to this title may be expended for any project in any area which is within the "Appalachian region" (as that term is defined in section 403 of the Appalachian Regional Development Act of 1965) which is approved for assistance under the Appalachian Regional Development Act of 1965. Ante, p. 21.

SEC. 105. There is hereby authorized to be appropriated to carry out this title not to exceed $500,000,000 for the fiscal year ending June 30, 1966, and for each fiscal year thereafter through the fiscal year ending June 30, 1969. Appropriation.

FINANCIAL ASSISTANCE FOR SEWER FACILITIES

SEC. 106. No financial assistance, through grants, loans, guarantees, or otherwise, shall be made under this Act to be used directly or indirectly for sewer or other waste disposal facilities unless the Secretary of Health, Education, and Welfare certifies to the Secretary that any waste material carried by such facilities will be adequately treated before it is discharged into any public waterway so as to meet applicable Federal, State, interstate, or local water quality standards.

TITLE II—OTHER FINANCIAL ASSISTANCE

PUBLIC WORKS AND DEVELOPMENT FACILITY LOANS

SEC. 201. (a) Upon the application of any State, or political subdivision thereof, Indian tribe, or private or public nonprofit organization or association representing any redevelopment area or part thereof, the Secretary is authorized to purchase evidence of indebtedness and to make loans to assist in financing the purchase or development of land

and improvements for public works, public service, or development facility usage, including public works, public service, and development facility usage, to be provided by agencies of the Federal Government pursuant to legislation requiring that non-Federal entities bear some part of the cost thereof, and the acquisition, construction, rehabilitation, alteration, expansion, or improvement of such facilities, including related machinery and equipment, within a redevelopment area, if he finds that—

(1) the project for which financial assistance is sought will directly or indirectly—

(A) tend to improve the opportunities, in the area where such project is or will be located, for the successful establishment or expansion of industrial or commercial plants or facilities,

(B) otherwise assist in the creation of additional long-term employment opportunities for such area, or

(C) primarily benefit the long-term unemployed and members of low-income families or otherwise substantially further the objectives of the Economic Opportunity Act of 1964;

78 Stat. 508. 42 USC 2701 note.

(2) the funds requested for such project are not otherwise available from private lenders or from other Federal agencies on terms which in the opinion of the Secretary will permit the accomplishment of the project;

(3) the amount of the loan plus the amount of other available funds for such project are adequate to insure the completion thereof;

(4) there is a reasonable expectation of repayment; and

(5) such area has an approved overall economic development program as provided in section 202(b)(10) and the project for which financial assistance is sought is consistent with such program.

(b) Subject to section 701(5), no loan, including renewals or extensions thereof, shall be made under this section for a period exceeding forty years, and no evidence of indebtedness maturing more than forty years from the date of purchase shall be purchased under this section. Such loans shall bear interest at a rate not less than a rate determined by the Secretary of the Treasury taking into consideration the current average market yield on outstanding marketable obligations of the United States with remaining periods to maturity comparable to the average maturities of such loans, adjusted to the nearest one-eighth of 1 per centum, less not to exceed one-half of 1 per centum per annum.

(c) There are hereby authorized to be appropriated such sums as may be necessary to carry out the provisions of this section and section 202: *Provided*, That annual appropriations for the purpose of purchasing evidences of indebtedness, making and participating in loans, and guaranteeing loans shall not exceed $170,000,000, for the fiscal year ending June 30, 1966, and for each fiscal year thereafter through the fiscal year ending June 30, 1970.

(d) Except for projects specifically authorized by Congress, no financial assistance shall be extended under this section with respect to any public service or development facility which would compete with an existing privately owned public utility rendering a service to the public at rates or charges subject to regulation by a State or Federal regulatory body, unless the State or Federal regulatory body determines that in the area to be served by the facility for which the financial assistance is to be extended there is a need for an increase in such service (taking into consideration reasonably foreseeable future needs) which the existing public utility is not able to meet through its existing facilities or through an expansion which it agrees to undertake.

(e) The Secretary shall prescribe regulations which will assure that appropriate local governmental authorities have been given a reasonable opportunity to review and comment upon proposed projects under this section.

LOANS AND GUARANTEES

SEC. 202. (a) The Secretary is authorized (1) to purchase evidences of indebtedness and to make loans (which for purposes of this section shall include participations in loans) to aid in financing any project within a redevelopment area for the purchase or development of land and facilities (including machinery and equipment) for industrial or commercial usage, including the construction of new buildings, the rehabilitation of abandoned or unoccupied buildings, and the alteration, conversion, or enlargement of existing buildings; and (2) to guarantee loans for working capital made to private borrowers by private lending institutions in connection with projects in redevelopment areas assisted under subsection (a)(1) hereof, upon application of such institution and upon such terms and conditions as the Secretary may prescribe: *Provided, however,* That no such guarantee shall at any time exceed 90 per centum of the amount of the outstanding unpaid balance of such loan.

(b) Financial assistance under this section shall be on such terms and conditions as the Secretary determines, subject, however, to the following restrictions and limitations:

(1) Such financial assistance shall not be extended to assist establishments relocating from one area to another or to assist subcontractors whose purpose is to divest, or whose economic success is dependent upon divesting, other contractors or subcontractors of contracts theretofore customarily performed by them: *Provided, however,* That such limitation shall not be construed to prohibit assistance for the expansion of an existing business entity through the establishment of a new branch, affiliate, or subsidiary of such entity if the Secretary finds that the establishment of such branch, affiliate, or subsidiary will not result in an increase in unemployment of the area of original location or in any other area where such entity conducts business operations, unless the Secretary has reason to believe that such branch, affiliate, or subsidiary is being established with the intention of closing down the operations of the existing business entity in the area of its original location or in any other area where it conducts such operations.

(2) Such assistance shall be extended only to applicants, both private and public (including Indian tribes), which have been approved for such assistance by an agency or instrumentality of the State or political subdivision thereof in which the project to be financed is located, and which agency or instrumentality is directly concerned with problems of economic development in such State or subdivision.

(3) The project for which financial assistance is sought must be reasonably calculated to provide more than a temporary alleviation of unemployment or underemployment within the redevelopment area wherein it is or will be located.

(4) No loan or guarantee shall be extended hereunder unless the financial assistance applied for is not otherwise available from private lenders or from other Federal agencies on terms which in the opinion of the Secretary will permit the accomplishment of the project.

(5) The Secretary shall not make any loan without a participation unless he determines that the loan cannot be made on a participation basis.

(6) No evidences of indebtedness shall be purchased and no loans shall be made or guaranteed unless it is determined that there is reasonable assurance of repayment.

(7) Subject to section 701(5) of this Act, no loan, including renewals or extension thereof, may be made hereunder for a period exceeding twenty-five years and no evidences of indebtedness maturing more than twenty-five years from date of purchase may be purchased hereunder: *Provided*, That the foregoing restrictions on maturities shall not apply to securities or obligations received by the Secretary as a claimant in bankruptcy or equitable reorganization or as a creditor in other proceedings attendant upon insolvency of the obligor.

(8) Loans made and evidences of indebtedness purchased under this section shall bear interest at a rate not less than a rate determined by the Secretary of the Treasury taking into consideration the current average market yield on outstanding marketable obligations of the United States with remaining periods to maturity comparable to the average maturities of such loans, adjusted to the nearest one-eighth of 1 per centum, plus additional charge, if any, toward covering other costs of the program as the Secretary may determine to be consistent with its purpose.

(9) Loan assistance shall not exceed 65 per centum of the aggregate cost to the applicant (excluding all other Federal aid in connection with the undertaking) of acquiring or developing land and facilities (including machinery and equipment), and of constructing, altering, converting, rehabilitating, or enlarging the building or buildings of the particular project, and shall, among others, be on the condition that—

(A) other funds are available in an amount which, together with the assistance provided hereunder, shall be sufficient to pay such aggregate cost;

(B) not less than 15 per centum of such aggregate cost be supplied as equity capital or as a loan repayable in no shorter period of time and at no faster an amortization rate than the Federal financial assistance extended under this section is being repaid, and if such a loan is secured, its security shall be subordinate and inferior to the lien or liens securing such Federal financial assistance: *Provided, however*, That, except in projects involving financial participation by Indian tribes, not less than 5 per centum of such aggregate cost shall be supplied by the State or any agency, instrumentality, or political subdivision thereof, or by a community or area organization which is nongovernmental in character, unless the Secretary shall determine in accordance with objective standards promulgated by regulation that all or part of such funds are not reasonably available to the project because of the economic distress of the area or for other good cause, in which case he may waive the requirement of this provision to the extent of such unavailability, and allow the funds required by this subsection to be supplied by the applicant or by such other non-Federal source as may reasonably be available to the project;

(C) to the extent the Secretary finds such action necessary to encourage financial participation in a particular project by other lenders and investors, and except as otherwise provided in subparagraph (B), any Federal financial assistance extended under this section may be repayable only after other loans made in connection with such project have been repaid in full, and the security, if any, for such Federal financial assistance may be subordinate and inferior to the lien or liens securing other loans made in connection with the same project.

(10) No such assistance shall be extended unless there shall be submitted to and approved by the Secretary an overall program for the economic development of the area and a finding by the State, or any agency, instrumentality, or local political subdivision thereof, that the project for which financial assistance is sought is consistent with such program: *Provided*, That nothing in this Act shall authorize financial assistance for any project prohibited by laws of the State or local political subdivision in which the project would be located, nor prevent the Secretary from requiring such periodic revisions of previously approved overall economic development programs as he may deem appropriate.

ECONOMIC DEVELOPMENT REVOLVING FUND

SEC. 203. Funds obtained by the Secretary under section 201, loan funds obtained under section 403, and collections and repayments received under this Act, shall be deposited in an economic development revolving fund (hereinafter referred to as the "fund"), which is hereby established in the Treasury of the United States, and which shall be available to the Secretary for the purpose of extending financial assistance under sections 201, 202, and 403, and for the payment of all obligations and expenditures arising in connection therewith. There shall also be credited to the fund such funds as have been paid into the area redevelopment fund or may be received from obligations outstanding under the Area Redevelopment Act. The fund shall pay into miscellaneous receipts of the Treasury, following the close of each fiscal year, interest on the amount of loans outstanding under this Act computed in such manner and at such rate as may be determined by the Secretary of the Treasury taking into consideration the current average market yield on outstanding marketable obligations of the United States with remaining periods to maturity comparable to the average maturities of such loans, adjusted to the nearest one-eighth of 1 per centum, during the month of June preceding the fiscal year in which the loans were made.

75 Stat. 47.
42 USC 2501 note.

TITLE III—TECHNICAL ASSISTANCE, RESEARCH, AND INFORMATION

SEC. 301. (a) In carrying out his duties under this Act the Secretary is authorized to provide technical assistance which would be useful in alleviating or preventing conditions of excessive unemployment or underemployment (1) to areas which he has designated as redevelopment areas under this Act, and (2) to other areas which he finds have substantial need for such assistance. Such assistance shall include project planning and feasibility studies, management and operational assistance, and studies evaluating the needs of, and developing potentialities for, economic growth of such areas. Such assistance may be provided by the Secretary through members of his staff, through the payment of funds authorized for this section to other departments or agencies of the Federal Government, through the employment of private individuals, partnerships, firms, corporations, or suitable institutions, under contracts entered into for such purposes, or through grants-in-aid to appropriate public or private nonprofit State, area, district, or local organizations. The Secretary, in his discretion, may require the repayment of assistance provided under this subsection and prescribe the terms and conditions of such repayment.

(b) The Secretary is authorized to make grants to defray not to exceed 75 per centum of the administrative expenses of organizations which he determines to be qualified to receive grants-in-aid under subsection (a) hereof. In determining the amount of the non-Federal share of such costs or expenses, the Secretary shall give due consideration to all contributions both in cash and in kind, fairly evaluated, including but not limited to space, equipment, and services. Where practicable, grants-in-aid authorized under this subsection shall be used in conjunction with other available planning grants, such as urban planning grants authorized under the Housing Act of 1954, as amended, and highway planning and research grants authorized under the Federal Aid Highway Act of 1962, to assure adequate and effective planning and economical use of funds.

68 Stat. 590.
12 USC 1703 note.
76 Stat. 1145.
23 USC 101 note.

(c) To assist in the long-range accomplishment of the purposes of this Act, the Secretary, in cooperation with other agencies having similar functions, shall establish and conduct a continuing program of study, training, and research to (A) assist in determining the causes of unemployment, underemployment, underdevelopment, and chronic depression in the various areas and regions of the Nation, (B) assist in the formulation and implementation of national, State, and local programs which will raise income levels and otherwise produce solutions to the problems resulting from these conditions, and (C) assist in providing the personnel needed to conduct such programs. The program of study, training, and research may be conducted by the Secretary through members of this staff, through payment of funds authorized for this section to other departments or agencies of the Federal Government, or through the employment of private individuals, partnerships, firms, corporations, or suitable institutions, under contracts entered into for such purposes, or through grants to such individuals, organizations, or institutions, or through conferences and similar meetings organized for such purposes. The Secretary shall make available to interested individuals and organizations the results of such research. The Secretary shall include in his annual report under section 707 a detailed statement concerning the study and research conducted under this section together with his findings resulting therefrom and his recommendations for legislative and other action.

(d) The Secretary shall aid redevelopment areas and other areas by furnishing to interested individuals, communities, industries, and enterprises within such areas any assistance, technical information, market research, or other forms of assistance, information, or advice which would be useful in alleviating or preventing conditions of excessive unemployment or underemployment within such areas. The Secretary may furnish the procurement divisions of the various departments, agencies, and other instrumentalities of the Federal Government with a list containing the names and addresses of business firms which are located in redevelopment areas and which are desirous of obtaining Government contracts for the furnishing of supplies or services, and designating the supplies and services such firms are engaged in providing.

(e) The Secretary shall establish an independent study board consisting of governmental and nongovernmental experts to investigate the effects of Government procurement, scientific, technical, and other related policies, upon regional economic development. Any Federal officer or employee may, with the consent of the head of the department or agency in which he is employed, serve as a member of such board, but shall receive no additional compensation for such service. Other members of such board may be compensated in accordance with the provisions of section 701(10). The board shall report its findings, together with recommendations for the better coordination of such policies, to the Secretary, who shall transmit the report to the Congress not later than two years after the enactment of this Act.

Report to Congress.

SEC. 302. There is hereby authorized to be appropriated $25,000,000 annually for the purposes of this title, for the fiscal year ending June 30, 1966, and for each fiscal year thereafter through the fiscal year ending June 30, 1970. Appropriation

TITLE IV—AREA AND DISTRICT ELIGIBILITY

PART A—REDEVELOPMENT AREAS

AREA ELIGIBILITY

SEC. 401. (a) The Secretary shall designate as "redevelopment areas"—

(1) those areas in which he determines, upon the basis of standards generally comparable with those set forth in paragraphs (A) and (B), that there has existed substantial and persistent unemployment for an extended period of time and those areas in which he determines there has been a substantial loss of population due to lack of employment opportunity. There shall be included among the areas so designated any area—

(A) where the Secretary of Labor finds that the current rate of unemployment, as determined by appropriate annual statistics for the most recent available calendar year, is 6 per centum or more and has averaged at least 6 per centum for the qualifying time periods specified in paragraph (B); and

(B) where the Secretary of Labor finds that the annual average rate of unemployment has been at least—

(i) 50 per centum above the national average for three of the preceding four calendar years, or

(ii) 75 per centum above the national average for two of the preceding three calendar years, or

(iii) 100 per centum above the national average for one of the preceding two calendar years.

The Secretary of Labor shall find the facts and provide the data to be used by the Secretary in making the determinations required by this subsection;

(2) those additional areas which have a median family income not in excess of 40 per centum of the national median, as determined by the most recent available statistics for such areas;

(3) those additional Federal or State Indian reservations or trust or restricted Indian-owned land areas which the Secretary, after consultation with the Secretary of the Interior or an appropriate State agency, determines manifest the greatest degree of economic distress on the basis of unemployment and income statistics and other appropriate evidence of economic underdevelopment;

(4) upon request of such areas, those additional areas in which the Secretary determines that the loss, removal, curtailment, or closing of a major source of employment has caused within three years prior to, or threatens to cause within three years after, the date of the request an unusual and abrupt rise in unemployment of such magnitude that the unemployment rate for the area at the time of the request exceeds the national average, or can reasonably be expected to exceed the national average, by 50 per centum or more unless assistance is provided. Notwithstanding any provision of subsection 401(b) to the contrary, an area designated under the authority of this paragraph may be given a reasonable time after designation in which to submit the overall economic development program required by subsection 202(b)(10) of this Act;

(5) notwithstanding any provision of this section to the contrary, those additional areas which were designated redevelopment areas under the Area Redevelopment Act on or after March 1, 1965: *Provided, however*, That the continued eligibility of such areas after the first annual review of eligibility conducted in accordance with section 402 of this Act shall be dependent on their qualification for designation under the standards of economic need set forth in subsections (a) (1) through (a) (4) of this section.

75 Stat. 47.
42 USC 2501 note.

(b) The size and boundaries of redevelopment areas shall be as determined by the Secretary: *Provided, however*, That—

(1) no area shall be designated until it has an approved overall economic development program in accordance with subsection 202(b) (10) of this Act;

(2) any area which does not submit an acceptable overall economic development program in accordance with subsection 202 (b) (10) of this Act within a reasonable time after notification of eligibility for designation, shall not thereafter be designated prior to the next annual review of eligibility in accordance with section 402 of this Act;

(3) no area shall be designated which does not have a population of at least one thousand five hundred persons, except for areas designated under subsection 401 (a) (3), which shall have a population of not less than one thousand persons; and

(4) except for areas designated under subsections (a) (3) and (a) (4) hereof, no area shall be designated which is smaller than a "labor area" (as defined by the Secretary of Labor), a county, or a municipality with a population of over two hundred and fifty thousand, whichever in the opinion of the Secretary is appropriate.

(c) Upon the request of the Secretary, the Secretary of Labor, the Secretary of Agriculture, the Secretary of the Interior, and such other heads of agencies as may be appropriate are authorized to conduct such special studies, obtain such information, and compile and furnish to the Secretary such data as the Secretary may deem necessary or proper to enable him to make the determinations provided for in this section. The Secretary shall reimburse when appropriate, out of any funds appropriated to carry out the purposes of this Act, the foregoing officers for any expenditures incurred by them under this section.

(d) If a State has no area designated under the preceding subsections of this section as a redevelopment area, the Secretary shall designate as a redevelopment area that area in such State which in his opinion most nearly qualifies under such preceding subsections. An area so designated shall have its eligibility terminated in accordance with the provisions of section 402 if any other area within the same State subsequently has become qualified or been designated under any other subsection of this section as of the time of the annual review prescribed by section 402: *Provided*, That the Secretary shall not terminate any designation of an area in a State as a redevelopment area if to do so would result in such State having no redevelopment area.

"Redevelopment area."

(e) As used in this Act, the term "redevelopment area" refers to any area within the United States which has been designated by the Secretary as a redevelopment area.

ANNUAL REVIEW OF AREA ELIGIBILITY

SEC. 402. The Secretary shall conduct an annual review of all areas designated in accordance with section 401 of this Act, and on the basis thereof shall terminate or modify the designations of such areas in accordance with objective standards which he shall prescribe by regulation. No area previously designated shall retain its designated status unless it maintains a currently approved overall economic development program in accordance with subsection 202(b)(10). No termination of eligibility shall (1) be made without thirty days' prior notification to the area concerned, (2) affect the validity of any application filed, or contract or undertaking entered into, with respect to such area pursuant to this Act prior to such termination, (3) prevent any such area from again being designated a redevelopment area under section 401 of this Act if the Secretary determines it to be eligible under such section, or (4) be made in the case of any designated area where the Secretary determines that an improvement in the unemployment rate of a designated area is primarily the result of increased employment in occupations not likely to be permanent. The Secretary shall keep the departments and agencies of the Federal Government, and interested State or local agencies, advised at all times of any changes made hereunder with respect to the classification of any area.

PART B—ECONOMIC DEVELOPMENT DISTRICTS

SEC. 403. (a) In order that economic development projects of broader geographical significance may be planned and carried out, the Secretary is authorized—

(1) to designate appropriate "economic development districts" within the United States with the concurrence of the States in which such districts will be wholly or partially located, if—

(A) the proposed district is of sufficient size or population, and contains sufficient resources, to foster economic development on a scale involving more than a single redevelopment area;

(B) the proposed district contains two or more redevelopment areas;

(C) the proposed district contains one or more redevelopment areas or economic development centers identified in an approved district overall economic development program as having sufficient size and potential to foster the economic growth activities necessary to alleviate the distress of the redevelopment areas within the district; and

(D) the proposed district has a district overall economic development program which includes adequate land use and transportation planning and contains a specific program for district cooperation, self-help, and public investment and is approved by the State or States affected and by the Secretary;

(2) to designate as "economic development centers," in accordance with such regulations as he shall prescribe, such areas as he may deem appropriate, if—

(A) the proposed center has been identified and included in an approved district overall economic development program and recommended by the State or States affected for such special designation;

(B) the proposed center is geographically and economically so related to the district that its economic growth may reasonably be expected to contribute significantly to the alleviation of distress in the redevelopment areas of the district; and

(C) the proposed center does not have a population in excess of two hundred and fifty thousand according to the last preceding Federal census.

(3) to provide financial assistance in accordance with the criteria of sections 101, 201, and 202 of this Act, except as may be

herein otherwise provided, for projects in economic development centers designated under subsection (a)(2) above, if—

(A) the project will further the objectives of the overall economic development program of the district in which it is to be located;

(B) the project will enhance the economic growth potential of the district or result in additional long-term employment opportunities commensurate with the amount of Federal financial assistance requested; and

(C) the amount of Federal financial assistance requested is reasonably related to the size, population, and economic needs of the district;

(4) subject to the 20 per centum non-Federal share required for any project by subsection 101(c) of this Act, to increase the amount of grant assistance authorized by section 101 for projects within redevelopment areas (designated under section 401), by an amount not to exceed 10 per centum of the aggregate cost of any such project, in accordance with such regulations as he shall prescribe if—

(A) the redevelopment area is situated within a designated economic development district and is actively participating in the economic development activities of the district; and

(B) the project is consistent with an approved district overall economic development program.

(b) In designating economic development districts and approving district overall economic development programs under subsection (a) of this section, the Secretary is authorized, under regulations prescribed by him—

(1) to invite the several States to draw up proposed district boundaries and to identify potential economic development centers;

(2) to cooperate with the several States—

(A) in sponsoring and assisting district economic planning and development groups, and

(B) in assisting such district groups to formulate district overall economic development programs;

(3) to encourage participation by appropriate local governmental authorities in such economic development districts.

(c) The Secretary shall by regulation prescribe standards for the termination or modification of economic development districts and economic development centers designated under the authority of this section.

"Economic development district."

(d) As used in this Act, the term "economic development district" refers to any area within the United States composed of cooperating redevelopment areas and, where appropriate, designated economic development centers and neighboring counties or communities, which has been designated by the Secretary as an economic development district.

"Economic development center."

(e) As used in this Act, the term "economic development center" refers to any area within the United States which has been identified as an economic development center in an approved district overall economic development program and which has been designated by the Secretary as eligible for financial assistance under sections 101, 201, and 202 of this Act in accordance with the provisions of this section.

"Local government."

(f) For the purpose of this Act the term "local government" means any city, county, town, parish, village, or other general-purpose political subdivision of a State.

Appropriation.

(g) There is hereby authorized to be appropriated not to exceed $50,000,000 for the fiscal year ending June 30, 1967, and for each fiscal year thereafter through the fiscal year ending June 30, 1970, for financial assistance extended under the provisions of subsection (a) (3) and (a) (4) hereof.

(h) In order to allow time for adequate and careful district planning, subsection (g) of this section shall not be effective until one year from the date of enactment. Effective date.

TITLE V—REGIONAL ACTION PLANNING COMMISSIONS

ESTABLISHMENT OF REGIONS

SEC. 501. The Secretary is authorized to designate appropriate "economic development regions" within the United States with the concurrence of the States in which such regions will be wholly or partially located if he finds (A) that there is a relationship between the areas within such region geographically, culturally, historically, and economically, (B) that with the exception of Alaska and Hawaii, the region is within contiguous States, and (C) upon consideration of the following matters, among others, that the region has lagged behind the whole Nation in economic development:

(1) the rate of unemployment is substantially above the national rate;

(2) the median level of family income is significantly below the national median;

(3) the level of housing, health, and educational facilities is substantially below the national level;

(4) the economy of the area has traditionally been dominated by only one or two industries, which are in a state of long-term decline;

(5) the rate of outmigration of labor or capital or both is substantial;

(6) the area is adversely affected by changing industrial technology;

(7) the area is adversely affected by changes in national defense facilities or production; and

(8) indices of regional production indicate a growth rate substantially below the national average.

REGIONAL COMMISSIONS

SEC. 502. (a) Upon designation of development regions, the Secretary shall invite and encourage the States wholly or partially located within such regions to establish appropriate multistate regional commissions.

(b) Each such commission shall be composed of one Federal member, hereinafter referred to as the "Federal cochairman", appointed by the President by and with the advice and consent of the Senate, and one member from each participating State in the region. Each State member may be the Governor, or his designee, or such other person as may be provided by the law of the State which he represents. The State members of the commission shall elect a cochairman of the commission from among their number. Membership.

(c) Decisions by a regional commission shall require the affirmative vote of the Federal cochairman and of a majority, or at least one if only two, of the State members. In matters coming before a regional commission, the Federal cochairman shall, to the extent practicable, consult with the Federal departments and agencies having an interest in the subject matter.

(d) Each State member of a regional commission shall have an alternate, appointed by the Governor or as otherwise may be provided Alternates.

by the law of the State which he represents. The President, by and with the advice and consent of the Senate, shall appoint an alternate for the Federal cochairman of each regional commission. An alternate shall vote in the event of the absence, death, disability, removal, or resignation of the State of Federal cochairman for which he is an alternate.

Compensation.

78 Stat. 417.
5 USC 2211.

78 Stat. 400.
5 USC 1113.

(e) The Federal cochairman to a regional commission shall be compensated by the Federal Government from funds authorized by this Act up to level IV of the Federal Executive Salary Schedule. His alternate shall be compensated by the Federal Government from funds authorized by this Act at not to exceed the maximum scheduled rate for grade GS–18 of the Classification Act of 1949, as amended, and when not actively serving as an alternate for the Federal cochairman shall perform such functions and duties as are delegated to him by the Federal cochairman. Each State member and his alternate shall be compensated by the State which they represent at the rate established by the law of such State.

(f) If the Secretary finds that the State of Alaska or the State of Hawaii meet the requirements for an economic development region, he may establish a Commission for either State in a manner agreeable to him and to the Governor of the affected State.

FUNCTIONS OF COMMISSION

SEC. 503. (a) In carrying out the purposes of this Act, each Commission shall with respect to its region—

(1) advise and assist the Secretary in the identification of optimum boundaries for multistate economic development regions;

(2) initiate and coordinate the preparation of long-range overall economic development programs for such regions;

(3) foster surveys and studies to provide data required for the preparation of specific plans and programs for the development of such regions;

(4) advise and assist the Secretary and the States concerned in the initiation and coordination of economic development districts, in order to promote maximum benefits from the expenditure of Federal, State, and local funds;

(5) promote increased private investment in such regions;

(6) prepare legislative and other recommendations with respect to both short-range and long-range programs and projects for Federal, State, and local agencies;

(7) develop, on a continuing basis, comprehensive and coordinated plans and programs and establish priorities thereunder, giving due consideration to other Federal, State, and local planning in the region;

(8) conduct and sponsor investigations, research, and studies, including an inventory and analysis of the resources of the region, and, in cooperation with Federal, State and local agencies, sponsor demonstration projects designed to foster regional productivity and growth;

(9) review and study, in cooperation with the agency involved, Federal, State, and local public and private programs and, where appropriate, recommend modifications or additions which will increase their effectiveness in the region;

(10) formulate and recommend, where appropriate, interstate compacts and other forms of interstate cooperation, and work with State and local agencies in developing appropriate model legislation; and

(11) provide a forum for consideration of problems of the region and proposed solutions and establish and utilize, as appropriate, citizens and special advisory councils and public conferences.

(b) The Secretary shall present such plans and proposals of the commissions as may be transmitted and recommended to him (but are not authorized by any other section of this Act) first for review by the Federal agencies primarily interested in such plans and proposals and then, together with the recommendations of such agencies, to the President for such action as he may deem desirable.

(c) The Secretary shall provide effective and continuing liaison between the Federal Government and each regional commission.

(d) Each Federal agency shall, consonant with law and within the limits of available funds, cooperate with such commissions as may be established in order to assist them in carrying out their functions under this section.

(e) Each regional commission may, from time to time, make additional recommendations to the Secretary and recommendations to the State Governors and appropriate local officials, with respect to—

(1) the expenditure of funds by Federal, State, and local departments and agencies in its region in the fields of natural resources, agriculture, education, training, health and welfare, transportation, and other fields related to the purposes of this Act; and

(2) such additional Federal, State, and local legislation or administrative actions as the commission deems necessary to further the purposes of this Act.

PROGRAM DEVELOPMENT CRITERIA

SEC. 504. In developing recommendations for programs and projects for future regional economic development, and in establishing within those recommendations a priority ranking for such programs and projects, the Secretary shall encourage each regional commission to follow procedures that will insure consideration of the following factors:

(1) the relationship of the project or class of projects to overall regional development including its location in an area determined by the State to have a significant potential for growth;

(2) the population and area to be served by the project or class of projects including the relative per capita income and the unemployment rates in the area;

(3) the relative financial resources available to the State or political subdivisions or instrumentalities thereof which seek to undertake the project;

(4) the importance of the project or class of projects in relation to other projects or classes of projects which may be in competition for the same funds;

(5) the prospects that the project, on a continuing rather than a temporary basis, will improve the opportunities for employment, the average level of income, or the economic and social development of the area served by the project.

REGIONAL TECHNICAL AND PLANNING ASSISTANCE

SEC. 505. (a) The Secretary is authorized to provide to the commissions technical assistance which would be useful in aiding the commissions to carry out their functions under this Act and to develop recommendations and programs. Such assistance shall include studies and plans evaluating the needs of, and developing potentialities for, economic growth of such region, and research on improving the conservation and utilization of the human and natural resources of the region. Such assistance may be provided by the Secretary through members of his staff, through the payment of funds authorized for this section to other departments or agencies of the Federal Government, or through the employment of private individuals, partnerships, firms, corporations, or suitable institutions, under contracts entered into for such purposes, or through grants-in-aid to the commissions. The Secretary, in his discretion, may require the repayment of assistance provided under this subsection and prescribe the terms and conditions of such repayment.

Federal share of costs.

(b) For the period ending on June 30 of the second full Federal fiscal year following the date of establishment of a commission, the administrative expenses of each commission as approved by the Secretary shall be paid by the Federal Government. Thereafter, not to exceed 50 per centum of such expenses may be paid by the Federal Government. In determining the amount of the non-Federal share of such costs or expenses, the Secretary shall give due consideration to all contributions both in cash and in kind, fairly evaluated, including but not limited to space, equipment, and services.

Appropriation.

(c) There is hereby authorized to be appropriated $15,000,000 for the fiscal year ending June 30, 1966, and for each fiscal year thereafter through the fiscal year ending June 30, 1970, for the purposes of this section.

ADMINISTRATIVE POWERS OF REGIONAL COMMISSIONS

SEC. 506. To carry out its duties under this Act, each regional commission is authorized to—

(1) adopt, amend, and repeal bylaws, rules, and regulations governing the conduct of its business and the performance of its functions;

(2) appoint and fix the compensation of an executive director and such other personnel as may be necessary to enable the commission to carry out its functions, except that such compensation shall not exceed the salary of the alternate to the Federal cochairman on the commission and no member, alternate, officer, or employee of such commission, other than the Federal cochairman on the commission and his staff and his alternate, and Federal employees detailed to the commission under clause (3), shall be deemed a Federal employee for any purpose;

(3) request the head of any Federal department or agency (who is hereby so authorized) to detail to temporary duty with the commission such personnel within his administrative jurisdiction as the commission may need for carrying out its functions, each such detail to be without loss of seniority, pay, or other employee status;

(4) arrange for the services of personnel from any State or local government or any subdivision or agency thereof, or any intergovernmental agency;

(5) make arrangements, including contracts, with any participating State government for inclusion in a suitable retirement and employee benefit system of such of its personnel as may not be eligible for, or continue in, another governmental retirement or employee benefit system, or otherwise provide for such coverage of its personnel, and the Civil Service Commission of the United States is authorized to contract with such commission for continued coverage of commission employees, who at date of

commission employment are Federal employees, in the retirement program and other employee benefit programs of the Federal Government;

(6) accept, use, and dispose of gifts or donations of services or property, real, personal, or mixed, tangible or intangible;

(7) enter into and perform such contracts, leases, cooperative agreements, or other transactions as may be necessary in carrying out its functions and on such terms as it may deem appropriate, with any department, agency, or instrumentality of the United States or with any State, or any political subdivision, agency, or instrumentality thereof, or with any person, firm, association, or corporation;

(8) maintain an office in the District of Columbia and establish field offices at such other places as it may deem appropriate; and

(9) take such other actions and incur such other expenses as may be necessary or appropriate.

INFORMATION

SEC. 507. In order to obtain information needed to carry out its duties, each regional commission shall—

(1) hold such hearings, sit and act at such times and places, take such testimony, receive such evidence, and print or otherwise reproduce and distribute so much of its proceedings and reports thereon as it may deem advisable, a cochairman of such commission, or any member of the commission designated by the commission for the purpose, being hereby authorized to administer oaths when it is determined by the commission that testimony shall be taken or evidence received under oath;

(2) arrange for the head of any Federal, State, or local department or agency (who is hereby so authorized, to the extent not otherwise prohibited by law) to furnish to such commission such information as may be available to or procurable by such department or agency; and

(3) keep accurate and complete records of its doings and transactions which shall be made available for public inspection. Records.

PERSONAL FINANCIAL INTERESTS

SEC. 508. (a) Except as permitted by subsection (b) hereof, no State member or alternate and no officer or employee of a regional commission shall participate personally and substantially as member, alternate, officer, or employee, through decision, approval, disapproval, recommendation, the rendering of advice, investigation, or otherwise, in any proceeding, application, request for a ruling or other determination, contract, claim, controversy, or other particular matter in which, to his knowledge, he, his spouse, minor child, partner, organization (other than a State or political subdivision thereof) in which he is serving as officer, director, trustee, partner, or employee, or any person or organization with whom he is serving as officer, director, trustee, partner, or employee, or any person or organization with whom he is negotiating or has any arrangement concerning prospective employment, has a financial interest. Any person who shall violate the provisions of this subsection shall be fined not more than $10,000, or imprisoned not more than two years, or both. Penalty.

(b) Subsection (a) hereof shall not apply if the State member, alternate, officer, or employee first advises the regional commission involved of the nature and circumstances of the proceeding, application, request for a ruling or other determination, contract, claim, controversy, or other particular matter and makes full disclosure of the financial interest and receives in advance a written determination made by such commission that the interest is not so substantial as to be deemed likely to affect the integrity of the services which the commission may expect from such State member, alternate, officer, or employee.

(c) No State member of a regional commission, or his alternate, shall receive any salary, or any contribution to or supplementation of salary for his services on such commission from any source other than his State. No person detailed to serve a regional commission under authority of clause (4) of section 506 shall receive any salary or any contribution to or supplementation of salary for his services on such commission from any source other than the State, local, or intergovernmental department or agency from which he was detailed or from such commission. Any person who shall violate the provisions of this subsection shall be fined not more than $5,000, or imprisoned not more than one year, or both.

Penalty.

Conflict-of-interest.

(d) Notwithstanding any other subsection of this section, the Federal cochairman and his alternate on a regional commission and any Federal officers or employees detailed to duty with it pursuant to clause (3) of section 10 shall not be subject to any such subsection but shall remain subject to sections 202 through 209 of title 18, United States Code.

(e) A regional commission may, in its discretion, declare void and rescind any contract or other agreement pursuant to the Act in relation to which it finds that there has been a violation of subsection (a) or (c) of this section, or any of the provisions of sections 202 through 209, title 18, United States Code.

76 Stat. 1121.

ANNUAL REPORTS

SEC. 509. Each regional commission established pursuant to this Act shall make a comprehensive and detailed annual report each fiscal year to the Congress with respect to such commission's activities and recommendations for programs. The first such report shall be made for the first fiscal year in which such commission is in existence for more than three months. Such reports shall be printed and transmitted to the Congress not later than January 31 of the calendar year following the fiscal year with respect to which the report is made.

TITLE VI—ADMINISTRATION

SEC. 601. (a) The Secretary shall administer this Act and, with the assistance of an Assistant Secretary of Commerce, in addition to those already provided for, shall supervise and direct the Administrator created herein, and coordinate the Federal cochairmen appointed heretofore or subsequent to this Act. The Assistant Secretary created by this section shall be appointed by the President by and with the advice and consent of the Senate and shall be compensated at the rate provided for level IV of the Federal Executive Salary Schedule. Such Assistant Secretary shall perform such functions as the Secretary may prescribe. There shall be appointed by the President, by and with the advice and consent of the Senate, an Administrator for Economic Development who shall be compensated at the rate provided for level V of the Federal Executive Salary Schedule who shall perform such duties as are assigned by the Secretary.

Assistant Secretary of Commerce, appointment.

78 Stat. 417.
5 USC 2211.

Administrator for Economic Development, appointment.

(b) Paragraph (12) of subsection (d) of section 303 of the Federal Executive Salary Act of 1964 is amended by striking out "(4)" and inserting in lieu thereof "(5)".

(c) Subsection (e) of section 303 of the Federal Executive Salary Act of 1964 is amended by adding at the end thereof the following new paragraph: 78 Stat. 417. 5 USC 2211.

"(100) Administrator for Economic Development."

ADVISORY COMMITTEE ON REGIONAL ECONOMIC DEVELOPMENT

SEC. 602. The Secretary shall appoint a National Public Advisory Committee on Regional Economic Development which shall consist of twenty-five members and shall be composed of representatives of labor, management, agriculture, State and local governments, and the public in general. From the members appointed to such Committee the Secretary shall designate a Chairman. Such Committee, or any duly established subcommittee thereof, shall from time to time make recommendations to the Secretary relative to the carrying out of his duties under this Act. Such Committee shall hold not less than two meetings during each calendar year.

CONSULTATION WITH OTHER PERSONS AND AGENCIES

SEC. 603. (a) The Secretary is authorized from time to time to call together and confer with any persons, including representatives of labor, management, agriculture, and government, who can assist in meeting the problems of area and regional unemployment or underemployment.

(b) The Secretary may make provision for such consultation with interested departments and agencies as he may deem appropriate in the performance of the functions vested in him by this Act.

TITLE VII—MISCELLANEOUS

POWERS OF SECRETARY

SEC. 701. In performing his duties under this Act, the Secretary is authorized to—

(1) adopt, alter, and use a seal, which shall be judicially noticed;

(2) hold such hearings, sit and act at such times and places, and take such testimony, as he may deem advisable;

(3) request directly from any executive department, bureau, agency, board, commission, office, independent establishment, or instrumentality information, suggestions, estimates, and statistics needed to carry out the purposes of this Act; and each department, bureau, agency, board, commission, office, establishment or instrumentality is authorized to furnish such information, suggestions, estimates, and statistics directly to the Secretary;

(4) under regulations prescribed by him, assign or sell at public or private sale, or otherwise dispose of for cash or credit, in his discretion and upon such terms and conditions and for such consideration as he shall determine to be reasonable, any evidence of debt, contract, claim, personal property, or security assigned to or held by him in connection with loans made or evidences of indebtedness purchased under this Act, and collect or compromise all obligations assigned to or held by him in connection with such loans or evidences of indebtedness until such time as such obligations may be referred to the Attorney General for suit or collection;

(5) further extend the maturity of or renew any loan made or evidence of indebtedness purchased under this Act, beyond the periods stated in such loan or evidence of indebtedness or in this

Act, for additional periods not to exceed ten years, if such extension or renewal will aid in the orderly liquidation of such loan or evidence of indebtedness;

(6) deal with, complete, renovate, improve, modernize, insure, rent, or sell for cash or credit, upon such terms and conditions and for such consideration as he shall determine to be reasonable, any real or personal property conveyed to, or otherwise acquired by, him in connection with loans made or evidences of indebtedness purchased under this Act;

(7) pursue to final collection, by way of compromise or other administrative action, prior to reference to the Attorney General, all claims against third parties assigned to him in connection with loans made or evidences of indebtedness purchased under this Act. This shall include authority to obtain deficiency judgments or otherwise in the case of mortgages assigned to the Secretary. Section 3709 of the Revised Statutes, as amended (41 U.S.C. 5), shall not apply to any contract of hazard insurance or to any purchase or contract for services or supplies on account of property obtained by the Secretary as a result of loans made or evidences of indebtedness purchased under this Act if the premium therefor or the amount thereof does not exceed $1,000. The power to convey and to execute, in the name of the Secretary, deeds of conveyance, deeds of release, assignments and satisfactions of mortgages, and any other written instrument relating to real or personal property or any interest therein acquired by the Secretary pursuant to the provisions of this Act may be exercised by the Secretary or by any officer or agent appointed by him for that purpose without the execution of any express delegation of power or power of attorney;

(8) acquire, in any lawful manner, any property (real, personal, or mixed, tangible or intangible), whenever deemed necessary or appropriate to the conduct of the activities authorized in sections 201, 202, 301, 403, and 503 of this Act;

(9) in addition to any powers, functions, privileges, and immunities otherwise vested in him, take any and all actions, including the procurement of the services of attorneys by contract, determined by him to be necessary or desirable in making, purchasing, servicing, compromising, modifying, liquidating, or otherwise administratively dealing with or realizing on loans made or evidences of indebtedness purchased under this Act;

(10) employ experts and consultants or organizations therefor as authorized by section 15 of the Administrative Expenses Act of 1946 (5 U.S.C. 55a), compensate individuals so employed at rates not in excess of $100 per diem, including travel time, and allow them, while away from their homes or regular places of business, travel expenses (including per diem in lieu of subsistence) as authorized by section 5 of such Act (5 U.S.C. 73b–2) for persons in the Government service employed intermittently, while so employed: *Provided, however*, That contracts for such employment may be renewed annually;

60 Stat. 810.

60 Stat. 808; 75 Stat. 339, 340.

(11) sue and be sued in any court of record of a State having general jurisdiction or in any United States district court, and jurisdiction is conferred upon such district court to determine such controversies without regard to the amount in controversy; but no attachment, injunction, garnishment, or other similar process, mesne or final, shall be issued against the Secretary or his property.

Nothing herein shall be construed to except the activities under this Act from the application of sections 507(b) and 2679 of title 28, United States Code, and of section 367 of the Revised Statutes (5 U.S.C. 316); and 62 Stat. 910, 984; 75 Stat. 539.

(12) establish such rules, regulations, and procedures as he may deem appropriate in carrying out the provisions of this Act.

PREVENTION OF UNFAIR COMPETITION

SEC. 702. No financial assistance under this Act shall be extended to any project when the result would be to increase the production of goods, materials, or commodities, or the availability of services or facilities, when there is not sufficient demand for such goods, material, commodities, services, or facilities, to employ the efficient capacity of existing competitive commercial or industrial enterprises.

SAVING PROVISIONS

SEC. 703. (a) No suit, action, or other proceeding lawfully commenced by or against the Administrator or any other officer of the Area Redevelopment Administration in his official capacity or in relation to the discharge of his official duties under the Area Redevelopment Act, shall abate by reason of the taking effect of the provisions of this Act, but the court may, on motion or supplemental petition filed at any time within twelve months after such taking effect, showing a necessity for the survival of such suit, action, or other proceeding to obtain a settlement of the questions involved, allow the same to be maintained by or against the Secretary or the Administrator or such other officer of the Department of Commerce as may be appropriate. 75 Stat. 47. 42 USC 2501 note.

(b) Except as may be otherwise expressly provided in this Act, all powers and authorities conferred by this Act shall be cumulative and additional to and not in derogation of any powers and authorities otherwise existing. All rules, regulations, orders, authorizations, delegations, or other actions duly issued, made, or taken by or pursuant to applicable law, prior to the effective date of this Act, by any agency, officer, or office pertaining to any functions, powers, and duties under the Area Redevelopment Act shall continue in full force and effect after the effective date of this Act until modified or rescinded by the Secretary or such other officer of the Department of Commerce as, in accordance with applicable law, may be appropriate.

TRANSFER OF FUNCTIONS, EFFECTIVE DATE, AND LIMITATIONS ON ASSISTANCE

SEC. 704. (a) The functions, powers, duties, and authorities and the assets, funds, contracts, loans, liabilities, commitments, authorizations, allocations, and records which are vested in or authorized to be transferred to the Secretary of the Treasury under section 29(b) of the Area Redevelopment Act, and all functions, powers, duties, and authorities under section 29(c) of the Area Redevelopment Act are hereby vested in the Secretary. 42 USC 2525.

(b) The President may designate a person to act as Administrator under this Act until the office is filled as provided in this Act or until the expiration of the first period of sixty days following the effective date of this Act, whichever shall first occur. While so acting such person shall receive compensation at the rate provided by this Act for such office. Acting Administrator, designation.

Effective date.

(c) The provisions of this Act shall take effect upon enactment unless herein explicitly otherwise provided.

Pending project applications.

(d) Notwithstanding any requirements of this Act relating to the eligibility of areas, projects for which applications are pending before the Area Redevelopment Administration on the effective date of this Act shall for a period of one year thereafter be eligible for consideration by the Secretary for such assistance under the provisions of this Act as he may determine to be appropriate.

(e) No financial assistance authorized under this Act shall be used to finance the cost of facilities for the generation, transmission, or distribution of electric energy, except on projects specifically authorized by the Congress, or to finance the cost of facilities for the production or transmission of gas (natural, manufactured, or mixed).

SEPARABILITY

SEC. 705. Notwithstanding any other evidence of the intent of Congress, it is hereby declared to be the intent of Congress that if any provision of this Act or the application thereof to any persons or circumstances shall be adjudged by any court of competent jurisdiction to be invalid, such judgment shall not affect, impair, or invalidate the remainder of this Act or its application to other persons and circumstances, but shall be confined in its operation to the provision of this Act or the application thereof to the persons and circumstances directly involved in the controversy in which such judgment shall have been rendered.

APPLICATION OF ACT

SEC. 706. As used in this Act, the terms "State", "States", and "United States" include the several States, the District of Columbia, the Commonwealth of Puerto Rico, the Virgin Islands, Guam, and American Samoa.

ANNUAL REPORT

SEC. 707. The Secretary shall make a comprehensive and detailed annual report to the Congress of his operations under this Act for each fiscal year beginning with the fiscal year ending June 30, 1966. Such report shall be printed and shall be transmitted to the Congress not later than January 3 of the year following the fiscal year with respect to which such report is made.

USE OF OTHER FACILITIES

SEC. 708. (a) The Secretary is authorized to delegate to the heads of other departments and agencies of the Federal Government any of the Secretary's functions, powers, and duties under this Act as he may deem appropriate, and to authorize the redelegation of such functions, powers, and duties by the heads of such departments and agencies.

(b) Departments and agencies of the Federal Government shall exercise their powers, duties, and functions in such manner as will assist in carrying out the objectives of this Act.

(c) Funds authorized to be appropriated under this Act may be transferred between departments and agencies of the Government, if such funds are used for the purposes for which they are specifically authorized and appropriated.

SEC. 709. There are hereby authorized to be appropriated such sums as may be necessary to carry out those provisions of the Act for which specific authority for appropriations is not otherwise provided in this Act. Appropriations authorized under this Act shall remain available until expended unless otherwise provided by appropriations Acts.

PENALTIES

SEC. 710. (a) Whoever makes any statement knowing it to be false, or whoever willfully overvalues any security, for the purpose of obtaining for himself or for any applicant any financial assistance under section 101, 201, 202, or 403 or any extension thereof by renewal, deferment or action, or otherwise, or the acceptance, release, or substitution of security therefor, or for the purpose of influencing in any way the action of the Secretary, or for the purpose of obtaining money, property, or anything of value, under this Act, shall be punished by a fine of not more than $10,000 or by imprisonment for not more than five years, or both

(b) Whoever, being connected in any capacity with the Secretary, in the administration of this Act (1) embezzles, abstracts, purloins, or willfully misapplies any moneys, funds, securities, or other things of value, whether belonging to him or pledged or otherwise entrusted to him, or (2) with intent to defraud the Secretary or any other body politic or corporate, or any individual, or to deceive any officer, auditor, or examiner, makes any false entry in any book, report, or statement of or to the Secretary, or without being duly authorized draws any order or issues, puts forth, or assigns any note, debenture, bond, or other obligation, or draft, bill of exchange, mortgage, judgment, or decree thereof, or (3) with intent to defraud participates or shares in or receives directly or indirectly any money, profit, property, or benefit through any transaction, loan, grant, commission, contract, or any other act of the Secretary, or (4) gives any unauthorized information concerning any future action or plan of the Secretary which might affect the value of securities, or having such knowledge invests or speculates, directly or indirectly, in the securities or property of any company or corporation receiving loans, grants, or other assistance from the Secretary, shall be punished by a fine of not more than $10,000 or by imprisonment for not more than five years, or both.

EMPLOYMENT OF EXPEDITERS AND ADMINISTRATIVE EMPLOYEES

SEC. 711. No financial assistance shall be extended by the Secretary under section 101, 201, 202, or 403 to any business enterprise unless the owners, partners, or officers of such business enterprise (1) certify to the Secretary the names of any attorneys, agents, and other persons engaged by or on behalf of such business enterprise for the purpose of expediting applications made to the Secretary for assistance of any sort, under this Act, and the fees paid or to be paid to any such person; and (2) execute an agreement binding such business enterprise, for a period of two years after such assistance is rendered by the Secretary to such business enterprise, to refrain from employing, tendering any office or employment to, or retaining for professional services, any person who, on the date such assistance or any part thereof was rendered, or within one year prior thereto, shall have served as an officer, attorney, agent, or employee, occupying a position or engaging in activities which the Secretary shall have determined involve discretion with respect to the granting of assistance under this Act.

PREVAILING RATE OF WAGE AND FORTY-HOUR WEEK

SEC. 712. All laborers and mechanics employed by contractors or subcontractors on projects assisted by the Secretary under this Act shall be paid wages at rates not less than those prevailing on similar construction in the locality as determined by the Secretary of Labor in accordance with the Davis-Bacon Act, as amended (40 U.S.C. 276a—276a-5). The Secretary shall not extend any financial assistance under section 101, 201, 202, or 403 for such a project without first obtaining adequate assurance that these labor standards will be maintained upon the construction work. The Secretary of Labor shall have, with respect to the labor standards specified in this provision, the authority and functions set forth in Reorganization Plan Numbered 14 of 1950 (15 F.R. 3176; 64 Stat. 1267; 5 U.S.C. 133z-15), and section 2 of the Act of June 13, 1934, as amended (40 U.S.C 276c).

49 Stat. 1011; 78 Stat. 238.

63 Stat. 108.

RECORD OF APPLICATIONS

SEC. 713. The Secretary shall maintain as a permanent part of the records of the Department of Commerce a list of applications approved for financial assistance under section 101, 201, 202, or 403, which shall be kept available for public inspection during the regular business hours of the Department of Commerce. The following information shall be posted in such list as soon as each application is approved; (1) the name of the applicant and, in the case of corporate applications, the names of the officers and directors thereof, (2) the amount and duration of the loan or grant for which application is made, (3) the purposes for which the proceeds of the loan or grant are to be used, and (4) a general description of the security offered in the case of a loan.

RECORDS AND AUDIT

SEC. 714. (a) Each recipient of assistance under this Act shall keep such records as the Secretary shall prescribe, including records which fully disclose the amount and the disposition by such recipient of the proceeds of such assistance, the total cost of the project or undertaking in connection with which such assistance is given or used, and the amount and nature of that portion of the cost of the project or undertaking supplied by other sources, and such other records as will facilitate an effective audit.

(b) The Secretary and the Comptroller General of the United States, or any of their duly authorized representatives, shall have access for the purpose of audit and examination to any books, documents, papers, and records of the recipient that are pertinent to assistance received under this Act.

CONFORMING AMENDMENT

SEC. 715. All benefits heretofore specifically made available (and not subsequently revoked) under other Federal programs to persons or to public or private organizations, corporations, or entities in areas designated by the Secretary as "redevelopment areas" under section 5 of the Area Redevelopment Act, are hereby also extended, insofar as

75 Stat. 48. 42 USC 2504.

practicable, to such areas as may be designated as "redevelopment areas" or "economic development centers" under the authority of section 401 or 403 of this Act: *Provided, however*, That this section shall not be construed as limiting such administrative discretion as may have been conferred under any other law.

SEC. 716. All financial and technical assistance authorized under this Act shall be in addition to any Federal assistance previously authorized, and no provision hereof shall be construed as authorizing or permitting any reduction or diminution in the proportional amount of Federal assistance to which any State or other entity eligible under this Act would otherwise be entitled under the provisions of any other Act.

Approved August 26, 1965.

For Product Safety Concerns and Information please contact our EU representative GPSR@taylorandfrancis.com
Taylor & Francis Verlag GmbH, Kaufingerstraße 24, 80331 München, Germany

www.ingramcontent.com/pod-product-compliance
Ingram Content Group UK Ltd.
Pitfield, Milton Keynes, MK11 3LW, UK
UKHW022109190926
475702UK00013B/1346
9781138944800